T0348232

walkermaths 1.4

# Mathematical Reasoning

Charlotte Walker and Victoria Walker

Walker Maths: 1.4 Mathematical Reasoning
1st Edition
Charlotte Walker
Victoria Walker

Cover design: Cheryl Smith, Macarn Design
Text design: Cheryl Smith, Macarn Design
Production controller: Katie McCappin

Any URLs contained in this publication were checked for currency during the production process. Note, however, that the publisher cannot vouch for the ongoing currency of URLs.

Acknowledgements
The authors and publisher wish to thank Sarah Waters for the cover image. All other images are courtesy of Shutterstock and iStock.

We also wish to acknowledge the trusted kindred spirits throughout the country for your help in preparing this title.

© Cengage Learning Australia Pty Limited

Copyright Notice
Copyright: This book is not a photocopiable master. No part of the publication may be copied, stored or communicated in any form by any means (paper or digital), including recording or storing in an electronic retrieval system, without the written permission of the publisher. Education institutions that hold a current licence with Copyright Licensing New Zealand, may copy from this book in strict accordance with the terms of the CLNZ Licence.

For product information and technology assistance,
in Australia call **1300 790 853**;
in New Zealand call **0800 449 725**

For permission to use material from this text or product, please email **aust.permissions@cengage.com**

**National Library of New Zealand Cataloguing-in-Publication Data**
A catalogue record for this book is available from the National Library of New Zealand.

978 01 7047768 0

**Cengage Learning Australia**
Level 5, 80 Dorcas Street
Southbank VIC 3006 Australia

Printed in China by 1010 Printing International Limited.
3 4 5 6 7 27 26 25 24

This icon appears throughout the workbook. It indicates that there is a worksheet available which has been collated from the original Level 1 WalkerMaths series. The worksheets can be accessed via the Teacher Resource for this workbook (available for purchase from nz.sales@cengage.com). These worksheets are an additional resource which can be used to support your students throughout the teaching of this standard.

# CONTENTS

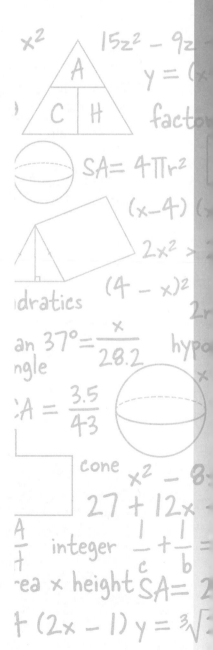

# Number

## Fundamentals

- These are some terms you should know.

### Factor

A factor divides into another number with no remainder.
The factors of **6** are **1**, **2**, **3**, **6**.
The factors of **12** are **1**, **2**, **3**, 4, **6**, 12.
The **common factors** of two numbers are factors of both.
The common factors of **6** and **12** are **1**, **2**, **3**, **6** because they are on both lists.
The **highest common factor** is the largest factor that is on both lists.
The highest common factor of **6** and **12** is **6**.

### Multiple

A multiple is the result of multiplying one number by another number.
The multiples of **3** are 3, **6**, 9, **12**, 15, **18**, …
The multiples of **2** are 2, 4, **6**, 8, 10, **12**, 14, 16, **18**, …
The **common multiples** of two numbers are multiples of both.
The common multiples of **3** and **2** are **6**, **12**, **18**, …
The **lowest common multiple** is the smallest multiple that is on both lists.
The lowest common multiple of **3** and **2** is **6**.

### Prime number

A prime number has exactly two factors: 1 and itself.
**Note: 1 is not a prime**, as it doesn't have exactly two factors.
**Examples:**

| Primes | 2, 3, 5, 7, 11, 13, … |
|---|---|
| Non-primes | 1, 4, 6, 8, 9, 10, 12, … |

### Standard form

Standard form is a way of writing either very large or very small numbers without writing all the place-holders (0s).
**Examples:**

| Ordinary form | Standard form |
|---|---|
| 12 000 | $1.2 \times 10^4$ |
| 7 | $7 \times 10^0$ |
| 0.0064 | $6.4 \times 10^{-3}$ |

PHOTOCOPYING OF THIS PAGE IS RESTRICTED UNDER LAW. ISBN: 9780170477680

## Square number

A square number is the result of multiplying an integer by itself.

**Example:** $4^2 = 16$, so 16 is a square number.

First six square numbers: 1, 4, 9, 16, 25, 36.

## Reciprocal

The reciprocal is 1 divided by a number or term.

**Examples:**

| Number or term | Reciprocal |
|:---:|:---:|
| 7 | $\frac{1}{7}$ |
| $\frac{2}{3}$ | $\frac{3}{2}$ |
| $4\frac{1}{2}$ | $\frac{2}{9}$ |
| $a$ | $\frac{1}{a}$ |

## Groups of numbers

*Real numbers (R)*

**Rational numbers (Q)**
Can be expressed as a fraction. Includes terminating and recurring decimals.

e.g. $\frac{3}{7}$, 1.49, $0.8\overset{\cdot\cdot}{1}3$

**Integers (I)**
..., −4, −3, −2, −1

**Whole numbers (W)**
0

**Natural or counting numbers (N)**
1, 2, 3, 4, ...

**Irrational numbers (Q')**
Non-terminating decimals including surds. e.g. $\sqrt{2}$, $\pi$

**Note:**

As well as the Real numbers, there are numbers that are known as Unreal or Imaginary numbers! (Only a pure mathematician could come up with such an idea.) Some of you will meet these in Year 13.

# Prime factorisation

- All non-prime numbers can be written as products of **prime factors**.
- Prime numbers and their factors are very important in cryptography (writing and using codes), especially for use in for cyber security.

**Examples:**

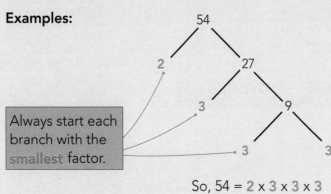

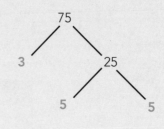

Always start each branch with the smallest factor.

So, 54 = **2** x **3** x **3** x **3**
or 54 = **2** x **3³**

75 = **3** x **5** x **5**
or 75 = **3** x **5²**

Complete these prime factor trees.

**1**

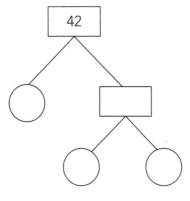

42 = _____

**2**

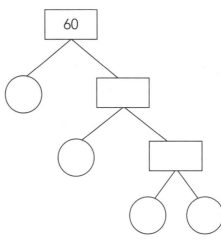

60 = _____

**3**

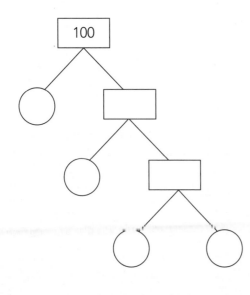

100 = _____

**4**

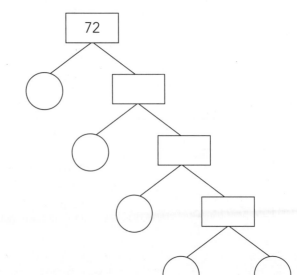

72 = _____

PHOTOCOPYING OF THIS PAGE IS RESTRICTED UNDER LAW.    ISBN: 9780170477680

# Ratio

- A ratio shows how an amount is split into several shares, usually of different sizes.
- Both parts of a ratio must use the **same units**.
- Ratios are generally written using whole numbers, not decimals or fractions, and are written in their simplest form.
- A **colon** (:) is used to separate the two numbers, e.g. 2:3 means 'two parts to three parts'.

## Finding a ratio

**Example:**

If the height and width of this rectangle are doubled, what is the ratio of the area of the new rectangle to the area of the original rectangle?

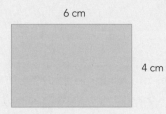

6 cm

4 cm

Area of the original rectangle = 6 x 4

= 24 cm²

The new rectangle = 12 cm x 8 cm

Area of the new rectangle = 12 x 8

= 96 cm²

The ratio of the area of the new rectangle to the area of the old rectangle is 96:24.
In its simplest form, this is 4:1.

## Finding a ratio where the total is given

**Example:**

ABCD is an 18 m straight line. AB:BC:CD = 3:2:1. What is the distance between B and C?

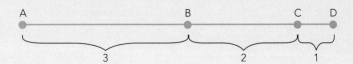

| A | B | C | D |

3        2        1

Step 1: Find the total number of parts:                              3 + 2 + 1 = 6

Step 2: Find the value of one part by dividing the total by the number of parts:        18 ÷ 6 = 3

Step 3: Multiply the value of one part by each part of the ratio:        3 x 3 : 2 x 3 : 1 x 3

= 9:6:3

The distance between B and C is 6 m.

## Finding a ratio where one part is given

**Example:**

The height of this triangle is 4 m. The ratio of the height of the triangle to its base is 2:5. Calculate the base of the triangle.

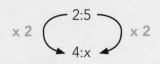

x 2    2:5    x 2

4:x

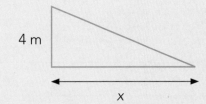

4 m

x

The base = x = 5 x 2

= 10 m

ISBN: 9780170477680    PHOTOCOPYING OF THIS PAGE IS RESTRICTED UNDER LAW.

Simplify these ratios.

**1**    16:24 = _____    **2**    20:32 = _____    **3**    65:100 = _____    **4**    25:350 = _____

**5**    Write the ratio 7:10 in the form $n$:1.    **6**    Write the ratio 9:4 in the form $n$:1.

_____        _____

Find the missing values.

**7**    9:27 = _____:3        **8**    8:7 = 40:_____

**9**    45:25 = 27:_____        **10**    7:21 = _____:33

Answer the following questions.

**11**    The height of a rectangle is 6 cm. The ratio of its height to its length is 2:3. Calculate the length of the rectangle.

**12**    Two sides of a triangle are XY = $6 \times 10^3$ cm and YZ = $3 \times 10^4$ cm. Write down the ratio XY:YZ in its simplest form.

**13**    The ratio of the angles in a triangle is $x$:$y$:$z$ = 5:4:3. Calculate the size of each angle.

**14**    Given that $\dfrac{a}{b} = \dfrac{3}{5}$ and $\dfrac{b}{c} = \dfrac{2}{3}$, find $a$:$b$:$c$.

**15**    If the height of this rectangle is tripled and the width is doubled, what is the ratio of the area of the new rectangle to the area of the original rectangle?

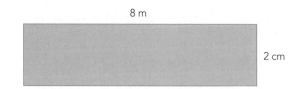

8 m

2 cm

**16**    The points A, B, C and D lie on a straight line.

AB:BD = 1:7

AC:CD = 13:3

Work out the ratios AB:BC:CD.

PHOTOCOPYING OF THIS PAGE IS RESTRICTED UNDER LAW.    ISBN: 9780170477680

**17** The width of the rectangle is 8 m. The ratio of the width of the rectangle to its height is 2:3. The ratio of the width of the rectangle to a side of the square is 4:5. Find the ratio of the area of the rectangle to the area of the square. Give the answer in its simplest form.

_____

_____

_____

8 m

**18** ABC is a straight line.
Angle CDB : angle ADB = 3:2.
Calculate the size of angle DBC.

_____

_____

_____

A

51°

B

74°

D          C

**19** The ratio of the height:length:depth of a cuboid is 1:4:3.
If its surface area is 237.5 cm$^2$, calculate its dimensions.

_____

_____

_____

_____

_____

height

depth

length

**20** The area of section A : area of section B is 1:6.
The radius of the small circle is 5 cm.
Calculate the radius of the large circle.

_____

_____

_____

_____

_____

B

45°

A

# Algebra

## Fundamentals

### Powers

- The words **power**, **exponent** and **index** (indices) all mean the same thing.

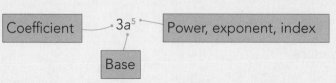

| Coefficient | $3a^5$ | Power, exponent, index |

Base

Remember:
- $a^4 = a \times a \times a \times a$
- $a^1 = a$
- $a^0 = 1$    As long as $a \neq 0$
- $a^{-3} = \dfrac{1}{a^3}$

- $-x^{\text{any power}} = $ a negative
- $(-x)^{\text{even power}} = $ a positive
- $(-x)^{\text{odd power}} = $ a negative

### Simplifying terms

#### Multiplying and dividing

| Unsimplified expressions | Simplified expressions |
|---|---|
| $a \times a \times a \times a$ | $a^4$ |
| $b \times -3b \times 5a$ | $-15ab^2$ |
| $a \div 6$ | $\dfrac{a}{6}$ or $\dfrac{1}{6}a$ |
| $\dfrac{27ab^2d}{9a^2bc}$ | $\dfrac{3bd}{ac}$ |

Because $1a$ and $1b$ are common to both the top and the bottom, they cancel out.

#### Adding and subtracting
- Terms can only be added or subtracted if they are 'like' terms.
- 'Like' terms must have exactly the same variables and each variable must be raised to exactly the same power.

#### Examples:
**1** $8a + a^2 - 3a = 5a + a^2$

**2** $7a^2bc - ab^2c - 2ab^2c = 7a^2bc - 3ab^2c$

PHOTOCOPYING OF THIS PAGE IS RESTRICTED UNDER LAW. ISBN: 9780170477680

## Expanding expressions

- In algebra, 'expand' means 'multiply out all the brackets'.
- After expanding, you are expected to collect the like terms in order to simplify the expression.
- The order in which you write your answer doesn't matter, as long as the signs for each term are correct.

**Examples:** Expand and simplify the following.

**1** $-2(7x + 6) = -14x - 12$

**2** $3x(6 - x) = 18x - 3x^2$

**3** $7 - 5(8x - 1) = 7 - 40x + 5$
$$= -40x + 12$$

**4** $4(3x - 2) - (6x + 5) = 12x - 8 - 6x - 5$
$$= 6x - 13$$

## Factorising expressions

- Factors are terms which are multiplied together (rather than added or subtracted), e.g. 2 and 3 are factors of 6 because 2 x 3 = 6.
- Expressions with brackets are usually in factorised form.

**Examples:**

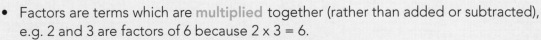

In each factorised expression, there is an unwritten x (times) sign before the bracket.

| Factorised form | Unfactorised (expanded) form |
|---|---|
| $2(4x + 5)$ | $8x + 10$ |
| $x(7 - 6x)$ | $7x - 6x^2$ |
| $4(3x^2 + x - 8)$ | $12x^2 + 4x - 32$ |

- When factorising, you must factorise completely. There must be no common factor for the terms inside the brackets.
- You need to ask yourself: 'What is the biggest thing (or things) that will divide into every term?'
- Do not use fractions when factorising.

**Examples:** Factorise the following.

The word 'fully' or 'completely' is understood, but not usually written.

**1** $12x - 28 = 4(3x - 7)$

**2** $24y^2 + 20y = 4y \times 6y + 4y \times 5$
$$= 4y(6y + 5)$$

All of the following are not fully factorised answers to **2**:

$4(6y^2 + 5y)$ $\qquad$ $2(12y^2 + 10y)$ $\qquad$ $2y(12y + 10)$ $\qquad$ $\frac{1}{2}(48y^2 + 40y)$

## Plotting linear equations

17

- There are several ways to plot linear equations.
- Tables will always work for any equation.
- Always plot **at least three points** and connect them with a **ruled line**.
- Always rule the line to the edges of the axes.

**Example:** Plot the line given by the equation $y + 2.5x = 12$.

Rearrange: $y = -2.5x + 12$

> It is easiest to complete the table if the equation is in the form $y = \ldots$

Fill in the table, plot the points, then connect them with a straight line.

| x | Calculation | y | Coordinates |
|---|---|---|---|
| 0 | −2.5 x 0 + 12 | 12 | (0, 12) |
| 2 | −2.5 x 2 + 12 | 7 | (2, 7) |
| 4 | −2.5 x 4 + 12 | 2 | (4, 2) |

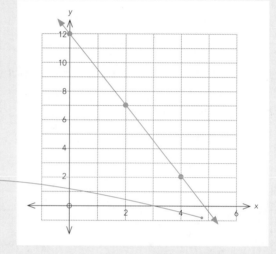

> Notice that line continues beyond the coordinates that you plotted.

## Using features of straight lines to write equations

18

- Graphs of lines are a way of illustrating relationships.
- The gradient and the **y** intercept enable us to write equations and interpret contexts.

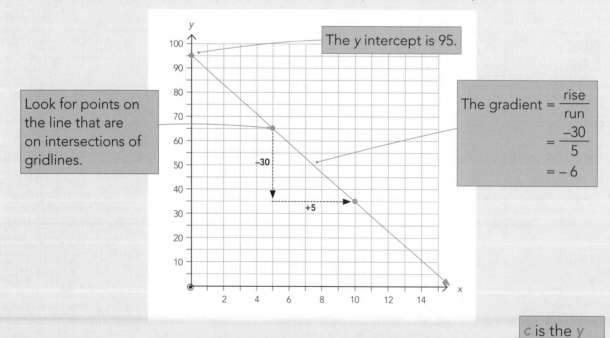

> Look for points on the line that are on intersections of gridlines.

> The y intercept is 95.

> The gradient $= \dfrac{\text{rise}}{\text{run}}$
> $= \dfrac{-30}{5}$
> $= -6$

To write an equation for this linear function, use the structure $y = mx + c$

> **c** is the **y** intercept.

So for this example, the equation is **$y = -6x + 95$**.

> **m** is the **gradient**.

PHOTOCOPYING OF THIS PAGE IS RESTRICTED UNDER LAW.   ISBN: 9780170477680

# Powers and roots

## Multiplying powers

When multiplying terms, we **add** the powers.

$$a^n \times a^m = a^{n+m}$$

Always deal with the coefficients first.

**Examples:**

1  $a^5 \times a^3 = a^8$    $5 + 3 = 8$

2  $4a^{3n}b \times 6a^nb^2 = 24a^{4n}b^3$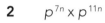

3  $a^{2n-1} \times a^{n+2} = a^{3n+1}$

4  $a^{1.3} \times a^{0.8} = a^{2.1}$

Simplify the following.

1  $d^2 \times d^4$

2  $p^{7n} \times p^{11n}$

3  $3y^2 \times 2y$

4  $e^4 \times e^3 \times e$

5  $5t \times 3t^{7+n} \times t^n$

6  $q \times q^{1.5}$

7  $-2w^6 \times w$

8  $z^{2.5} \times z^{0.6}$

9  $5a^{2d}b^6 \times 7a^3b^d$

10  $-4d \times (-3d^4)$

11  $-6k^3 \times 2k \times (-4k^{1.5})$

12  $v^3 \times (-11v^7) \times (3v^4)$

13  $p^{2.5} \times p^{0.6} \times p$

14  $a^{n-1} \times a \times a^{5n}$

ISBN: 9780170477680   PHOTOCOPYING OF THIS PAGE IS RESTRICTED UNDER LAW.

## Dividing powers

When dividing terms, we **subtract** the powers.

$$\frac{a^n}{a^m} = a^{n-m}$$

Again, deal with the coefficients first.

**Examples:**

**1** $b^{12} \div b^3$ or $\frac{b^{12}}{b^3} = b^9$   $12 - 3 = 9$

**2** $\frac{24x^{4n}}{6x^{7n}} = 4x^{4n-7n}$

$$= 4x^{-3n} \text{ or } \frac{4}{x^{3n}}$$

**3** $\frac{b^{2.4}}{b^{0.6}} = b^{1.8}$

**4** $\frac{a^8 b^3 c^6}{a^4 b^3 c^2} = a^4 c^4$

Simplify the following.

**1** $a^{15} \div a^5$

**2** $\frac{z^{12}}{z^4}$

**3** $d^{2m} \div d^{6m}$

**4** $\frac{18s^{8.5}}{3s}$

**5** $\frac{6z^6}{12z^{12}}$

**6** $-15m^{10} \div 3m^5$

**7** $(p^7 \times q^{3x}) \div (p^{2x} q)$

**8** $\frac{a^2 b^3 c^8}{ab^2 c^2}$

**9** $a^4 b^6 \div a^2 b^7$

**10** $\frac{27x^3 y^2}{3xy^5}$

**11** $\frac{j^2}{5j^5 k}$

**12** $-\frac{8y^7 z^2}{24y^2}$

**13** $\frac{12y^7 z}{24y^2 z^5}$

**14** $\frac{-10a^6 b^3 c}{-35ab^2 c^7}$

PHOTOCOPYING OF THIS PAGE IS RESTRICTED UNDER LAW.   ISBN: 9780170477680

## Powers of powers

When finding a power of a power, we **multiply** the powers.

$$(a^m)^n = a^{mn}$$

**Examples:**

**1** $(a^6)^3 = a^6 \times a^6 \times a^6$

    $= a^{18}$

Multiply the powers: 6 x 3 = 18

**2** $(4a^{3n})^2 = 4a^{3n} \times 4a^{3n}$

    $= 16a^{6n}$

**3** $\left(\dfrac{1}{2}a^4\right)^3 = \left(\dfrac{1}{2}\right)^3 \times (a^4)^3$

    $= \dfrac{1^3}{2^3} \times a^{4 \times 3}$

    $= \dfrac{1}{8}a^{12}$ or $\dfrac{a^{12}}{8}$

Simplify the following.

**1** $(y^{10})^3$

**2** $(2z^5)^4$

**3** $(-p^4)^2$

**4** $(-q^2)^5$

**5** $(-2x^{3n})^2$

**6** $(-3a^5)^3$

**7** $-2(4b^5c^7)^3$

**8** $(5p^b q^{3b} r^{2b})^3$

**9** $-(-4a^4bc^2)^3$

**10** $(2(-d^4)^2)^3$

**11** $\left(\dfrac{1}{3}x^5\right)^4$

**12** $\left(-\dfrac{4}{5}g^2\right)^3$

## Putting it together so far

**Examples:**

1   $\dfrac{2a^4 \times a^5}{a^2 \times a^3} = \dfrac{2a^9}{a^5}$

$= 2a^4$

2   $\dfrac{(4ab^3)^3}{2a^2} = \dfrac{64a^3b^9}{2a^2}$

$= 32ab^9$

3   $\dfrac{(2a^2)^3}{(3a^4)^2} = \dfrac{8a^6}{9a^8}$

$= \dfrac{8}{9a^2}$

4   $(b^4)^2(2b)^3 = b^8 \times 8 \times b^3$

$= 8b^{11}$

Simplify the following.

1   $(yz)^4 \div (y^2z^2)^2$

2   $\dfrac{(-4a)^2}{a^4}$

3   $\left(\dfrac{-x^2}{3}\right)^3$

4   $-\left(\dfrac{2ab}{10ab^4}\right)^2$

5   $(8b^3c^8)^2 \times 3bc$

6   $\dfrac{(3x^4)^3y}{xy^2 z}$

7   $\left(-\dfrac{1}{2}x^3\right)^2$

8   $\left(\dfrac{-x^2}{y^3}\right)^3$

9   $\dfrac{p^2q \times p^3q^4}{pq^2}$

10   $\dfrac{(3y^4z)^2 \times 5yz}{9y^2z^5}$

PHOTOCOPYING OF THIS PAGE IS RESTRICTED UNDER LAW.   ISBN: 9780170477680

## Expressions with a common base

**Steps:**   1   Rewrite the expression with the smallest base possible.

2   Use the power laws to simplify the expression.

Remember: $2^{n+1}$ can be written as $2^n \times 2^1$, which is $2(2^n)$.

**Examples:**

**1**   $64^x = (2^6)^x$
   $= 2^{6x}$

**2**   $125^{\frac{n}{3}} = (5^3)^{\frac{n}{3}}$
   $= 5^n$

**3**   $\dfrac{27^{2n}}{3^{4n}} = \dfrac{(3^3)^{2n}}{3^{4n}}$
   $= \dfrac{3^{6n}}{3^{4n}}$
   $= 3^{2n}$

**4**   $\dfrac{3^{n+1}}{9} = \dfrac{3^{n+1}}{3^2}$
   $= 3^{(n+1)-2}$
   $= 3^{n-1}$

Simplify the following.

**1**   $\dfrac{3^{2n}}{27}$

_____

_____

**2**   $16 \times 2^{3n}$

_____

_____

**3**   $\dfrac{81^n}{3^{4n}}$

_____

_____

**4**   $\dfrac{3^n \times 9^{2n}}{27^n}$

_____

_____

**5**   $\dfrac{5 \times 4^{3n}}{2^n}$

_____

_____

_____

**6**   $\left(\dfrac{3^{n+1}}{81}\right)^n$

_____

_____

_____

**7**   $\left(\dfrac{16^n}{8^{n-1}}\right)^{2n}$

_____

_____

**8**   $\dfrac{9^{n+2} \times 3^{5n}}{3^{4n-1}}$

_____

_____

ISBN: 9780170477680   PHOTOCOPYING OF THIS PAGE IS RESTRICTED UNDER LAW.

## Fractional powers and roots

- Finding a root is the reverse of finding a power.
- Roots can be written as fractional powers.

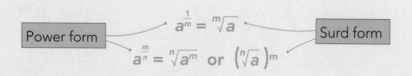

$$a^{\frac{1}{m}} = \sqrt[m]{a}$$

Power form — Surd form

$$a^{\frac{m}{n}} = \sqrt[n]{a^m} \text{ or } \left(\sqrt[n]{a}\right)^m$$

Notice that for a square root you can leave out the 2.

**Examples:**

**With numbers**

| Index form | Surd form |
| --- | --- |
| $25^{\frac{1}{2}}$ | $\sqrt[2]{25}$ or $\sqrt{25} = 5$ |
| $216^{\frac{1}{3}}$ | $\sqrt[3]{216} = 6$ |
| $17^{\frac{2}{3}}$ | $\sqrt[3]{17^2}$ or $\sqrt[3]{289}$ |
| $8^{\frac{2}{3}}$ | $\sqrt[3]{8^2} = \sqrt[3]{64}$ <br> $= 4$ |

**With variables**

| Index form | Surd form |
| --- | --- |
| $a^{\frac{2}{2}}$ | $\sqrt[2]{a^2}$ or $\sqrt{a^2}$ or $a$ |
| $a^{\frac{1}{2}}$ | $\sqrt[2]{a^1}$ or $\sqrt{a}$ |
| $a^{\frac{3}{4}}$ | $\sqrt[4]{a^3}$ or $\sqrt[4]{a}^3$ |
| $6a^{\frac{3}{5}}$ | $6\sqrt[5]{a^3}$ |

 23

**Write these in exponent form.**

**1** $\sqrt[4]{a}$ _____

**2** $\sqrt[3]{x^2}$ _____

**3** $9\sqrt{x}$ _____

**4** $5\sqrt[3]{y}$ _____

**5** $\sqrt{b^7}$ _____

**6** $11\sqrt[5]{a^2}$ _____

**Write these in surd form.**

**7** $x^{\frac{1}{3}}$ _____

**8** $y^{\frac{2}{5}}$ _____

**9** $7a^{\frac{1}{2}}$ _____

**10** $(7a)^{\frac{1}{2}}$ _____

**11** $3x^{\frac{4}{7}}$ _____

**12** $12b^{\frac{9}{2}}$ _____

PHOTOCOPYING OF THIS PAGE IS RESTRICTED UNDER LAW. ISBN: 9780170477680

## Negative powers

- These are the reciprocal of the positive power.
- To find the **reciprocal**, you turn the **fraction upside down**,

  e.g. the reciprocal of 2 (or $\frac{2}{1}$) is $\frac{1}{2}$.

**With numbers**

| Negative power | Reciprocal |
|---|---|
| $8^{-1}$ | $\frac{1}{8}$ |
| $\left(\frac{3}{4}\right)^{-1}$ | $\frac{4}{3}$ |
| $\left(\frac{3}{5}\right)^{-2}$ | $\frac{5^2}{3^2} = \frac{25}{9}$ |
| $(-3)^{-4}$ | $\frac{1^4}{3^4} = \frac{1}{81}$ |
| $\left(-\frac{6}{7}\right)^{-2}$ | $\frac{7^2}{6^2} = \frac{49}{36}$ |

**With variables**

| Negative power | Reciprocal |
|---|---|
| $x^{-2}$ | $\frac{1}{x^2}$ |
| $5y^{-3}$ | $5 \times \frac{1}{y^3} = \frac{5}{y^3}$ |
| $(5y)^{-3}$ | $\frac{1}{5^3 y^3} = \frac{1}{125 y^3}$ |
| $(2x^4 y)^{-3}$ | $\frac{1}{(2x^4 y)^3} = \frac{1}{8x^{12} y^3}$ |
| $3a^{-2} b^4 c^{-3}$ | $\frac{3b^4}{a^2 c^3}$ |

Simplify the following.

**1**   $4^{-1}$      _____

**2**   $\left(\frac{5}{8}\right)^{-1}$      _____

**3**   $a^{-3}$      _____

**4**   $xy^{-2}$      _____

**5**   $4x^{-7}$      _____

**6**   $x^{-4} y^2 z^{-3}$      _____

**7**   $(5b)^{-2}$      _____

**8**   $(xy)^{-5}$      _____

**9**   $(a^2 b)^{-3}$      _____

**10**   $2(x^3 y^2)^{-1}$      _____

**11**   $(4bc^3)^{-2}$      _____

**12**   $4(3x)^{-2}$      _____

**13**   $8(2a)^{-3}$      _____

**14**   $2(x^2 y^{-2})^{-2}$      _____

## Equations with powers

**Steps:**
1 Rewrite the expression with the smallest base possible.
2 Use the power laws to simplify the expression.
3 There must be **one** expression using the same base on each side,
   e.g. 2... = 2... or $y$... = $y$...

**Examples:** Find the values of $n$.

**1** $2^{3n} \times 2^{n-12} = 16$

$2^{3n+n-12} = 2^4$

$4n - 12 = 4$

$4n = 16$

$n = 4$

One expression using the **same base** on each side.

**2** $a^{4n-3} \times (a^n)^2 = a^9$

$a^{4n-3} \times a^{2n} = a^9$

$a^{6n-3} = a^9$

$6n - 3 = 9$

$6n = 12$

$n = 2$

If the base numbers are the same, then the powers must be equal.

Solve the following.

**1** $4^{3n} = 2^{n+5}$

**2** $(a^{2n-3})^2 = a^{3n} \times a^2$

**3** $6 \times 2^{n+1} = 192$

**4** $9 \times 3^{4n-1} = (3^n)^2$

**5** $\dfrac{4^{3n-1}}{2^n} = 2^3$

**6** $\dfrac{2^{2n+1}}{\sqrt{2}} = (2^n)^3$

**7** Write an equation for $p$ in terms of $q$.

$4 \times 2^{p+q} = 32^{2q}$

**8** Write an equation for $p$ in terms of $q$.

$(a^{q-2p})^2 = a \times a^{4-3p}$

PHOTOCOPYING OF THIS PAGE IS RESTRICTED UNDER LAW. ISBN: 9780170477680

# Algebraic fractions

## Multiplying and dividing algebraic fractions

### Multiplying fractions

$$\frac{a}{b} \times \frac{c}{d} = \frac{ac}{bd}$$

Multiply the numerators and multiply the denominators.

**Examples:**

1   $\dfrac{3z}{5} \times \dfrac{z}{4} = \dfrac{3z^2}{20}$

2   $\dfrac{2(b+1)}{6} \times \dfrac{4}{b} = \dfrac{8(b+1)}{6b} = \dfrac{4(b+1)}{3b}$

### Dividing fractions

- To find a **reciprocal**, you **invert** (turn upside down) its equivalent fraction.

**Examples:**

If a term is not a fraction, write it with a divisor of 1.

| Term | Reciprocal of term |
|---|---|
| $7 = \dfrac{7}{1}$ | $\dfrac{1}{7}$ |
| $a = \dfrac{a}{1}$ | $\dfrac{1}{a}$ |
| $2y = \dfrac{2y}{1}$ | $\dfrac{1}{2y}$ |
| $\dfrac{3a}{b}$ | $\dfrac{b}{3a}$ |

- To divide fractions, you **multiply** by the **reciprocal** of the **second** fraction.

$$\div \Rightarrow \times$$

$$\frac{a}{b} \div \frac{c}{d} = \frac{a}{b} \times \frac{d}{c} = \frac{ad}{bc}$$

Invert the second fraction ⇒ the reciprocal

**Examples:**

1   $a \div \dfrac{1}{a} = \dfrac{a}{1} \times \dfrac{a}{1}$

   $= a^2$

2   $\dfrac{30y}{z} \div \dfrac{1}{2} = \dfrac{30y}{z} \times \dfrac{2}{1}$

   $= \dfrac{60y}{z}$

3   $\dfrac{y}{3} \div \dfrac{y}{4} = \dfrac{y}{3} \times \dfrac{4}{y}$

   $= \dfrac{4y}{3y}$

   $= \dfrac{4}{3}$

4   $\dfrac{5b}{2} \div \dfrac{10}{b} = \dfrac{5b}{2} \times \dfrac{b}{10}$

   $= \dfrac{5b^2}{20}$

   $= \dfrac{b^2}{4}$

ISBN: 9780170477680 PHOTOCOPYING OF THIS PAGE IS RESTRICTED UNDER LAW.

Simplify the following. Do not expand the brackets.

1   $\dfrac{b}{5} \times \dfrac{3b}{6}$    _____

                 _____

2   $\dfrac{7}{y} \div \dfrac{4}{y}$    _____

                 _____

3   $\dfrac{6x}{y} \div \dfrac{2}{x}$    _____

                 _____

4   $\dfrac{3(b+1)}{2} \times \dfrac{b}{7}$    _____

                 _____

5   $\dfrac{12x^2}{3} \div x$    _____

                 _____

6   $\dfrac{a^2}{6} \times \dfrac{8b}{10a}$    _____

                 _____

7   $\dfrac{a}{2b^2} \div a$

_____

_____

8   $3y^2 \div \dfrac{z}{y}$

_____

_____

9   $\dfrac{-6(b-2)}{9} \times \dfrac{12}{5(b+4)}$

_____

_____

_____

10   $\dfrac{2(x-6)}{7} \div \dfrac{3(x-4)}{5}$

_____

_____

_____

11   $\dfrac{5y^2z}{z^3} \times \dfrac{1}{y} \times \dfrac{3z}{10}$

_____

_____

_____

12   $\dfrac{(3y)^2}{4z^3} \div \dfrac{y}{8z}$

_____

_____

_____

PHOTOCOPYING OF THIS PAGE IS RESTRICTED UNDER LAW.    ISBN: 9780170477680

## Adding and subtracting algebraic fractions

### 1 Where the denominators are the same

$$\frac{a}{b} + \frac{c}{b} = \frac{a+c}{b} \text{ or } \frac{a}{b} - \frac{c}{b} = \frac{a-c}{b}$$

Add or subtract the numerators.

The denominator stays the same.

**Examples:**

1 $\dfrac{b}{6} + \dfrac{3b}{6} = \dfrac{4b}{6}$

$\qquad = \dfrac{2b}{3}$

2 $\dfrac{5x}{9} - \dfrac{x-3}{9} = \dfrac{5x - (x-3)}{9}$

$\qquad = \dfrac{4x+3}{9}$  Watch the signs.

### 2 Where the denominators are different

Multiply both the numerator and the denominator by a number (or variable) that will create equal denominators.

$$\frac{a}{b} + \frac{c}{d} = \frac{a}{b} \times \frac{d}{d} + \frac{c}{d} \times \frac{b}{b}$$

$\dfrac{d}{d} = 1$ and $\dfrac{b}{b} = 1$

$$= \frac{ad}{bd} + \frac{cb}{db}$$

$$= \frac{ad + bc}{bd}$$

**Examples:**

1 $\dfrac{y}{5} + \dfrac{y}{3} = \dfrac{y}{5} \times \dfrac{3}{3} + \dfrac{y}{3} \times \dfrac{5}{5}$

$\qquad = \dfrac{3y}{15} + \dfrac{5y}{15}$

$\qquad = \dfrac{8y}{15}$

2 $\dfrac{x}{4} + x = \dfrac{x}{4} + \dfrac{x}{1} \times \dfrac{4}{4}$

$\qquad = \dfrac{x}{4} + \dfrac{4x}{4}$

$\qquad = \dfrac{5x}{4}$

3 $\dfrac{5x}{7} - \dfrac{2}{3} = \dfrac{5x}{7} \times \dfrac{3}{3} - \dfrac{2}{3} \times \dfrac{7}{7}$

$\qquad = \dfrac{15x}{21} - \dfrac{14}{21}$

$\qquad = \dfrac{15x - 14}{21}$

4 $\dfrac{4y+1}{3} - \dfrac{y}{2} = \dfrac{4y+1}{3} \times \dfrac{2}{2} - \dfrac{y}{2} \times \dfrac{3}{3}$

$\qquad = \dfrac{8y+2}{6} - \dfrac{3y}{6}$

$\qquad = \dfrac{5y+2}{6}$

Add and subtract the following. Give your answer in its simplest form.

1 $\dfrac{y}{8} + \dfrac{5y}{8}$

2 $\dfrac{a}{5} + \dfrac{b}{5}$

_____

_____

3 $\dfrac{9x}{4} - \dfrac{6x}{4}$

4 $\dfrac{2x^2}{3} - \dfrac{7}{3}$

_____

_____

ISBN: 9780170477680    PHOTOCOPYING OF THIS PAGE IS RESTRICTED UNDER LAW.

**5**  $\dfrac{5}{10} + \dfrac{3y}{10} - \dfrac{y}{10}$

_____

**6**  $\dfrac{8x}{11} + \dfrac{2x}{11}$

_____

**7**  $\dfrac{10x}{3} - \dfrac{x}{2}$

_____

_____

**8**  $\dfrac{4c}{5} + \dfrac{9}{10}$

_____

_____

**9**  $\dfrac{x+2}{5} + \dfrac{3x}{4}$

_____

_____

**10**  $\dfrac{5x-2}{7} - \dfrac{x}{3}$

_____

_____

**11**  $\dfrac{5x}{2} + x$

_____

_____

**12**  $\dfrac{4x}{5} + \dfrac{3x}{4}$

_____

_____

**13**  $\dfrac{7a}{9} - \dfrac{3a}{2} =$

_____

_____

**14**  $\dfrac{5ab}{8} - \dfrac{a}{3}$

_____

_____

**15**  $2 - \dfrac{x}{6}$

_____

_____

**16**  $\dfrac{2x+1}{5} + \dfrac{x}{4} =$

_____

_____

**17**  $\dfrac{3y-2}{4} + y =$

_____

_____

**18**  $y - \dfrac{4x+3}{7} =$

_____

_____

PHOTOCOPYING OF THIS PAGE IS RESTRICTED UNDER LAW.     ISBN: 9780170477680

## Simplifying algebraic fractions

- Where the denominator or numerator is a sum or a difference, **factorise**.
- 'Cancel' common factors. Remember: $\frac{a}{a} = 1$.

**Examples:** Simplify the following.

1  $\dfrac{3y}{6z-9} = \dfrac{\cancel{3}y}{\cancel{3}(2z-3)}$   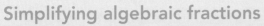 Cancel out 3.

$= \dfrac{y}{2z-3}$

2  $\dfrac{3ab^2 - 4a^3b + a^2b}{4ab^2} = \dfrac{ab(3b - 4a^2 + a)}{4ab^2}$   Cancel out $ab$.

$= \dfrac{3b - 4a^2 + a}{4b}$

28

Simplify the following.

1  $\dfrac{15x^2 + x}{xy}$   _____

2  $\dfrac{4p - p^2}{3p}$   _____

3  $\dfrac{a^2b - b}{ab}$   _____

4  $\dfrac{6b}{3a - 9b}$   _____

5  $\dfrac{x^3y}{3x^2 - xy}$   _____

6  $\dfrac{2ab^2}{a - 7ab}$   _____

7  $\dfrac{pq^3 + 3p^2q}{2pq}$   _____

8  $\dfrac{12pq - 3p}{6p^2q}$   _____

9  $\dfrac{2x^3y^2 - 3xy^2}{4x^2y + 5x^2y^3}$

10  $\dfrac{7x^3y^2 + 2x^2y - 3x^2}{x^2y^2}$

11  $\dfrac{15a^3b - 5ab^2 + 10b}{5b^2}$

12  $\dfrac{8x^3y^2z - 4xy^2}{12xyz + 4xy^2 - 8x^2y^2}$

# Substitution

- When substituting, you replace variables with numbers.
- It is important to remember BEDMAS when doing this.
- Don't forget that $\sqrt{x^2} = \pm x$.

**Examples:**

**1** Calculate the value of $W$ when $x = -3$ and $y = 45°$.

$W = x^2 + \tan y$

> Remember that a negative squared is a positive.

$W = (-3)^2 + \tan 45°$
$= 9 + 1$
$= 10$

**2** The surface area of a cylinder is $2\pi r^2 + 2\pi rh$. Calculate the value of the surface area when $r = 1.5$ cm and $h = 5$ cm. Leave $\pi$ in your answer.

$SA = 2\pi r^2 + 2\pi rh$
$= 2 \times \pi \times 1.5^2 + 2 \times \pi \times 1.5 \times 5$
$= 4.5\pi + 15\pi$
$= 19.5\pi$ cm$^2$

If $b = 2$, $c = -6$, $d = -3$ and $e = 4$, calculate the values of A.

**1** $A = c^2(b - e)$ _____

**2** $A = bd(c - e)$ _____

**3** $A = \dfrac{\sqrt{cd - b}}{e}$ _____

**4** $A = (b + c)(d - e)$ _____

**5** $A = c^2 + de - bd^2$ _____

**6** $A = \dfrac{c + \sqrt{ed^2}}{-b}$ _____

If $r = 1.5$ and $h = 5$, calculate the values of V. Leave $\pi$ in your answer.

**7** $V = \pi r^2 h$ _____

**8** $V = r^2 - \dfrac{1}{3}\pi r^3$ _____

If $x = -2$, $y = 30°$ and $z = 60°$, calculate the values of W.

**9** $W = \sin y - x^2$ _____

**10** $W = \dfrac{(\cos z)^2}{x}$ _____

PHOTOCOPYING OF THIS PAGE IS RESTRICTED UNDER LAW. ISBN: 9780170477680

# Solving equations

- 'Solve' means 'find a value for $x$'.

**Rules:**
1. You can do anything you like to an equation as long as you do the **same to both sides**.
2. There should be only **one equals sign** per line.
3. Collect all the variables on one side and numbers on the other side.
4. When you want to get rid of something, perform the **opposite** operation.
5. If the equation contains a root, isolate it to one side and then raise both sides to the relevant power.
6. Your answer should always be in the form $x = \ldots$

**Trick:** If you need to change the sign of everything, multiply **both** sides by $-1$.

**Examples:**

**1**
$$7(2x - 1) = 2(4 + 6x) \quad \text{(expand)}$$
$$14x - 7 = 8 + 12x \quad (-12x)$$
$$2x - 7 = 8 \quad (+7)$$
$$2x = 15 \quad (\div 2)$$
$$x = 7.5$$

**2**
$$\frac{4x - 1}{2} = \frac{3 + 5x}{3} \quad (\times 6)$$
$$3(4x - 1) = 2(3 + 5x) \quad \text{(expand)}$$
$$12x - 3 = 6 + 10x \quad (-10x)$$
$$2x - 3 = 6 \quad (+3)$$
$$2x = 9 \quad (\div 2)$$
$$x = 4.5$$

**3**
$$2\sqrt[3]{3x + 1} - 5 = 21 \quad (+5)$$
$$2\sqrt[3]{3x + 1} = 26 \quad (\div 2)$$
$$\sqrt[3]{3x + 1} = 13 \quad \text{(cube)}$$
$$3x + 1 = 13^3 \quad (-1)$$
$$3x = 2196 \quad (\div 3)$$
$$x = 732$$

**4**
$$\frac{4x}{7} - 3 = \frac{x}{2} \quad (\times \frac{14}{1})$$
$$\frac{4x}{7} \times \frac{14}{1} - 3 \times \frac{14}{1} = \frac{x}{2} \times \frac{14}{1}$$
$$8x - 42 = 7x \quad (-7x)$$
$$x - 42 = 0 \quad (+42)$$
$$x = 42$$

Solve the following.

**1** $\quad 5 - 2(x + 7) = 45$

**2** $\quad \dfrac{7 + x}{3} = \dfrac{5x + 2}{4}$

_____

_____

_____

**3** $\quad 3\sqrt{2x - 4} = 24$

**4** $\quad 9x^2 + 4 = 85$

_____

_____

_____

**5**   $4(x + 5) - 5 = -3(1 - 4x)$

**6**   $\dfrac{2x}{5} - 6 = 0.8$

**7**   $\dfrac{5x}{8} - 3 = \dfrac{x}{2}$

**8**   $\sqrt[3]{4(x + 5)} - 11 = 9$

**9**   $\dfrac{8x}{3} - \dfrac{2x + 5}{4} = -11$

**10**   $\dfrac{6(x + 7)}{2} + 5 = \dfrac{9x - 2}{5}$

**11**   $\dfrac{3}{4}x + 0.5 = 2x - \dfrac{1}{3}$

**12**   $\dfrac{3}{5}x - 7 = \dfrac{x}{4} + 0.35$

**13**   $\dfrac{1}{2}(4x - 10) - 2(x - 3) = \sqrt{x - 1}$

**14**   $\sqrt[3]{x^2 + 2} - 3 = 0$

PHOTOCOPYING OF THIS PAGE IS RESTRICTED UNDER LAW.   ISBN: 9780170477680

## Forming and solving linear equations

**1  Given relationships between angles, lengths, areas and volumes**
- Define your variable using the first letter of the word whose value you need to find, e.g. perimeter = $p$.
- If you need to find several values, it is usually easiest to select the smaller value as your variable.
- Often it is useful to draw a diagram.

**Words used for the four basic operations:**

| + | plus, total, more, and, add(ed), sum increased by, greater than | − | subtract(ed), less, decreased by |
|---|---|---|---|
| × | of, times, multiplied by, product | ÷ | divided by, shared between, out of, per |

- 'Double' and 'twice' both mean multiply by two.

**Example:**
The perimeter of this triangle is 96 cm.
Calculate the value of $x$.

$$2x + x + (x + 8) = 96$$
$$4x + 8 = 96$$
$$4x = 88$$
$$x = 22 \text{ cm}$$

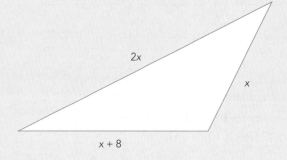

Form equations for these situations and then solve them.

**1**  The perimeter of this rectangle is 24 cm.
Find the value of $x$ and each side length.

_____

_____

_____

$4x - 3$

$x + 1$

**2**  Find the value of $x$ and the size of each interior angle of this triangle.

_____

_____

_____

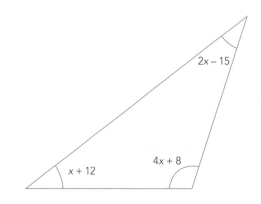

$2x - 15$

$4x + 8$

$x + 12$

ISBN: 9780170477680  PHOTOCOPYING OF THIS PAGE IS RESTRICTED UNDER LAW.

**3** The ends of three radii touch the circumference of a circle at the points a, b and c. Is bc a diameter of the circle?

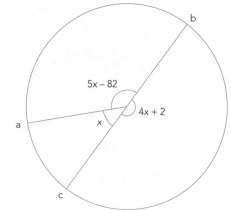

_____

_____

_____

_____

**4** Is the perimeter of this equilateral triangle greater than 150 cm?

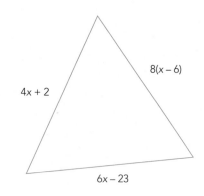

_____

_____

_____

_____

**5** The height of a rectangle is 80% larger than its width. If the perimeter is 784 cm, how long is each side?

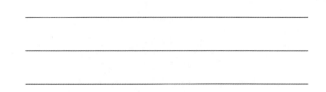

_____

_____

_____

_____

**6** The perimeter of the triangle is half the perimeter of the rectangle. Calculate the value of x.

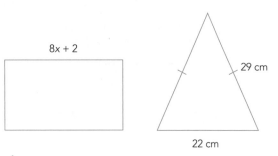

_____

_____

_____

_____

**7** Find the values of x and y in this parallelogram.

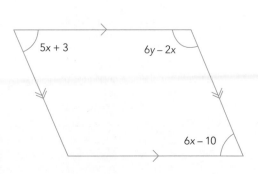

_____

_____

_____

_____

PHOTOCOPYING OF THIS PAGE IS RESTRICTED UNDER LAW.   ISBN: 9780170477680

**2 Given two pairs of values (points on the graph) that satisfy the equation**
- You might be given two points on a graph or two pairs of values for a linear relationship.
- In these cases you will need to calculate the gradient (m) of the line using the formula:

$$m = \frac{\text{rise}}{\text{run}} = \frac{y_2 - y_1}{x_2 - x_1}$$

- Then use the gradient (m) along with the coordinates of one point $(x_1, y_1)$ in the formula:

$$y - y_1 = m(x - x_1)$$

**Example:** Calculate the equation for the line that connects (–8, 6) and (2, –6).

Step 1: Name the coordinates:      (–8, 6)   and   (2, –6).

$$x_1 \quad y_1 \quad\quad x_2 \quad y_2$$

Step 2: Calculate the gradient:

$$m = \frac{y_2 - y_1}{x_2 - x_1}$$
$$= \frac{-6 - 6}{2 - (-8)}$$
$$= \frac{-12}{10}$$
$$= -1.2$$

Substitiute the gradient and one set of coordinates.

Step 3: Calculate the equation:

$$y - y_1 = m(x - x_1)$$
$$y - 6 = -1.2(x - (-8)) \quad \text{(expand and add 6)}$$
$$y = -1.2x - 9.6 + 6$$
$$y = -1.2x - 3.6$$

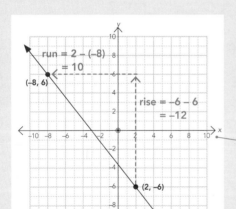

run = 2 – (–8)
= 10

(–8, 6)

rise = –6 – 6
= –12

(2, –6)

The calculation of the gradient is the same as you would do on a graph.

Calculate the equations for the lines that connect these points.

**8**    (2, 9) and (7, 14)

_____

_____

_____

_____

_____

**9**    (0, 5) and (2, 9)

_____

_____

_____

_____

_____

**10**  (2, 2) and (4, 8)

_____

_____

_____

_____

_____

**11**  (5, –7) and (–4, 11)

_____

_____

_____

_____

_____

**12**  (38, –8) and (–6, 14)

_____

_____

_____

_____

_____

**13**  (6, –8) and (36, 37)

_____

_____

_____

_____

_____

**14**  (6, –8) and (36, 37)

_____

_____

_____

_____

_____

**15**  (–10, –15) and (–35, 5)

_____

_____

_____

_____

_____

**16**  (4, –3p) and (9, 2p)

_____

_____

_____

_____

**17**  (2p, 8) and (–p, 17)

_____

_____

_____

_____

PHOTOCOPYING OF THIS PAGE IS RESTRICTED UNDER LAW.     ISBN: 9780170477680

### 3 Given that equations represent parallel or perpendicular lines

- **Parallel lines** have equal gradients. This means that $m_1 = m_2$.

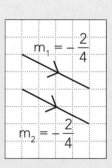

- **Perpendicular lines** are at right angles to each other and they have gradients which are negative reciprocals of each other. This means that $m_1 \times m_2 = -1$.

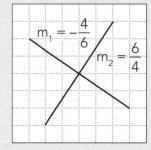

**Examples:**

**1** Write down the equation of a line which passes through the point (3, –5) and is parallel to $y = px - 7$.

Parallel $\Rightarrow m = p$

$$y - y_1 = m(x - x_1)$$
$$y - (-5) = p(x - 3)$$
$$y = px - 3p - 5$$

**2** Calculate the value of $p$ which would make these two lines perpendicular.

A:  $y = -2.5x + 7$
B:  $5y = px - 125$

From A:              $m_1 = -2.5$

Perpendicular $\Rightarrow m_1 \times m_2 = -1$

$$\therefore m_2 = \frac{-1}{-2.5} = 0.4$$

From B:          $y = \frac{p}{5}x - 25$

$$\therefore \frac{p}{5} = 0.4 \text{ so } p = 2$$

Calculate the equations for the lines which connect these points.

**18** A line passes through the origin, and is parallel to the line $3y = -x + 23$.
Write down its equation.

_____

_____

**19** Write down the equation of a line which passes through the point (12, –4) and is parallel to $y = \frac{2}{3}x + 8$.

_____

_____

**20** A line has a $y$ intercept of 5 and is perpendicular to the line $y = 4x - 9$.
Write down its equation.

_____

_____

_____

**21** A line has a $y$ intercept of –7 and is perpendicular to the line $5y = -6x - 11$.
Write down its equation.

_____

_____

_____

**22** Find the equation of a line which is parallel to $2y = 7x - 3$, and which passes through the point (4, 12).

_____

_____

_____

_____

_____

_____

**23** Find the equation of a line which is perpendicular to $8y - 3x + 15$, and which passes through the point (21, 6).

_____

_____

_____

_____

_____

_____

**24** The line $y = ax + 8$ is perpendicular to the line $2x + 6y - 7 = 0$. Find the value of a.

_____

_____

_____

_____

_____

_____

**25** The line $5x + by + 3 = 0$ is parallel to the line $2x + y - 5 = 0$. Find the value of b.

_____

_____

_____

_____

_____

_____

**26** Calculate the value of $p$ which would make these two lines parallel.
A:     $12y = 8x - 7$
B:     $-9y = px + 10$

_____

_____

_____

_____

_____

**27** Calculate the value of $p$ which would make these two lines perpendicular.
A:     $5y = -2x + 5$
B:     $15x = py + 18$

_____

_____

_____

_____

_____

PHOTOCOPYING OF THIS PAGE IS RESTRICTED UNDER LAW.     ISBN: 9780170477680

# Inequalities

## Understanding inequation signs

- Inequations are mathematical statements that have <, ≤, > or ≥ signs.

  < means 'is less than'
  > means 'is greater than'
  ≤ means 'is less than or equal to', 'is at most'
  ≥ means 'is greater than or equal to', 'is at least'

  > The bigger number is at the bigger end of the sign:
  > bigger > smaller
  > smallest ≤ biggest, etc.

- Sometimes there are two inequalities in a statement.
  e.g. $a < x < b$ means '$x$ is greater than $a$ and less than $b$'
  or '$x$ is between $a$ and $b$'.

  > The hollow dot shows that $x$ cannot be **exactly** –1.

**Examples:** Visualise these on a number line.

**1** $x < -1$ means $x$ is less than –1.

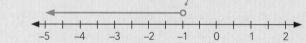

**2** $x ≥ 5$ means that $x$ is greater than or equal to five.

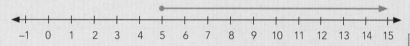

> The solid dot means that $x$ can equal 2.

**3** $-6 < x ≤ 2$ means $x$ is greater than –6 and less than or equal to 2.

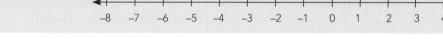

Place an inequality symbol between these pairs of values.

**1**  6 ☐ 11  **2**  4 ☐ –2  **3**  –9 ☐ –8  **4**  –0.5 ☐ 0

Write statements for these inequations.

**5**  $x < 9$

_____

**6**  $x < -2$

_____

**7**  $1 < x < 14$

_____

**8**  $x ≥ -7$

_____

**9**  $x ≤ 0$

_____

**10**  $5 ≤ x ≤ 198$

_____

Write inequations for these statements.

**11**  $x$ is less than 11. _____

**12**  $x$ is greater than or equal to 2. _____

**13**  $x$ is between –1 and 4 inclusive.

_____

**14**  $x$ is less than or equal to 7.

_____

**15**  $x$ is greater than or equal to 5 but less than 8. _____

## Using inequality expressions

- An inequality can be used to represent a situation where a measurement has been rounded to the nearest unit.

**Example:** A length is 4 m to the nearest metre.
This means it has a minimum length of 3.5 m and it must be **less than** 4.5 m.

| 3.5 rounded to the nearest metre is 4 m. | 4.5 rounded to the nearest metre is **5** m. |

It could be represented as: $3.5 \leq x < 4.5$ or $3.5 \leq x \leq 4.4\dot{9}$

Write statements for these.

**1** A and B are two **consecutive** odd numbers between 0 and 6 and $B > A$. Give an example of what these two numbers could be.

_____

**2** If $y = 3x$ and $x \leq 8$, what are the possible values of $y$?

_____

**3** This diagram is drawn to scale.
Which of these statements are true?

☐ $a + c > b$

☐ $a > c$

☐ $c < b$

☐ $a + b + c < 180°$

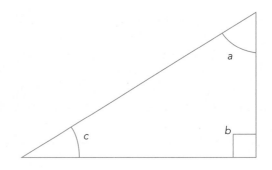

**4** $x$ is an integer and $-2 \leq x \leq 3$.

Tick the correct box for each statement.

Hint: Substitute the extremes, $-2$ and $3$, into the statement and see if it is true.

|  | Always true | Sometimes true | Never true |
|---|---|---|---|
| $4 < x$ | ☐ | ☐ | ☐ |
| $x + 1 < 0$ | ☐ | ☐ | ☐ |
| $x^2 > 10$ | ☐ | ☐ | ☐ |

PHOTOCOPYING OF THIS PAGE IS RESTRICTED UNDER LAW.    ISBN: 9780170477680

Write statements for these. Include units where appropriate.

**5**  **a**  A rectangle is 6 cm long and 2 cm high, both measured to the nearest centimetre. Complete the statements for its possible length and height.

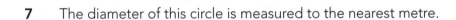

2 cm

6 cm

_____ ≤ L < _____          _____ ≤ H < _____

**b**  Calculate the range of values for the area of the rectangle.

_____ __ A __ _____

**c**  Calculate the range of values for the perimeter of the rectangle.

_____ __ P __ _____

**6**  Consider the statement $10 < (x + 3) < 25$. If $x$ is an integer:

**a**  What is the largest possible value of $x$? _____

**b**  What is the smallest possible value of $x$? _____

**c**  If the value of $x$ did not have to be an integer, would the greatest possible value of $x$ increase or decrease? _____

**7**  The diameter of this circle is measured to the nearest metre.

**a**  Complete this statement for the possible diameter.

_____ __ d __ _____

**b**  Complete this statement for the possible radius.

_____ __ r __ _____

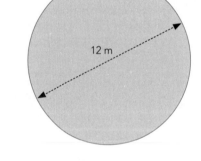

12 m

**c**  Complete this statement for the possible area, leaving pi in your answer.

_____ __ A __ _____

Write a sentence comparing these values to the area if the diameter of 12 m was used.

_____

**8**  If $5\ \text{cm} \le x \le 9\ \text{cm}$ and $2.5\ \text{cm} \le y \le 4.8\ \text{cm}$, complete this statement to show the range of areas of the ring, leaving pi in your answer.

_____ __ A __ _____

_____ __ A __ _____

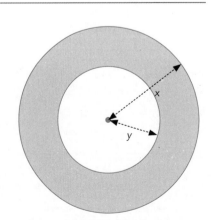

x

y

## Solving linear inequations

- You solve an inequation in exactly the same way as you solve an equation **except**:

    if you need to **multiply or divide** the equation by a **negative** number

  **or**  if you **swap the sides**,

    then you must **reverse the sign** (e.g. write 10 < 5x as 5x > 10).

**Examples:**

**1**
$$7x \geq 42 \qquad (\div\ 7)$$
$$x \geq \frac{42}{7}$$
$$x \geq 6$$

So x is greater than or equal to 6.

**2**
$$4x - 2 < -38 \qquad (+\ 2)$$
$$4x < -36 \qquad (\div\ 4)$$
$$x < -9$$

So x is less than −9.

**3**
$$41 < 5 + 3x \qquad (-\ 5)$$
$$36 < 3x \qquad \text{(swap sides)}$$
$$3x > 36 \qquad (\div\ 3)$$
$$x > 12$$

So x is greater than 12.

Swap the sides ⇒ reverse the sign.

**4**
$$\frac{x + 1}{2} < \frac{2 + 5x}{6} \qquad (\times\ 12)$$
$$6x + 6 < 4 + 10x \qquad (-\ 6)$$
$$6x < -2 + 10x \qquad (-\ 10x)$$
$$-4x < -2 \qquad (\div\ -4)$$
$$x > 0.5$$

So x is greater than 0.5.

Divide by a **negative** number ⇒ **reverse the sign.**

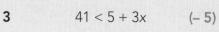

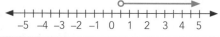

You can check the direction of the arrow by substituting x = 0 into the equation.

Solve the following and complete the sentence.

**1**  8x < 32

_____

_____

So x is _____ than _____

**2**  4x − 2 ≤ 50

_____

_____

So x is _____ than or equal to _____

**3**  5 + 9x > 77

_____

_____

So x is _____ than _____

**4**  11x − 1 ≥ x + 19

_____

_____

So x is _____

PHOTOCOPYING OF THIS PAGE IS RESTRICTED UNDER LAW.   ISBN: 9780170477680

**5** $\dfrac{5x}{7} < x + 6$

_____

_____

_____

So x is _____

**6** $x - 3 \le \dfrac{3x + 1}{4}$

_____

_____

_____

So x is _____

**7** $\dfrac{2(x - 1)}{10} \ge 2$

_____

_____

_____

So x is _____

**8** $\dfrac{6(x - 3)}{5} > 25.2$

_____

_____

_____

So x is _____

**9** $\dfrac{3x}{2} + \dfrac{x}{5} \le 17$

_____

_____

_____

So x is _____

**10** $\dfrac{2x - 3}{5} > \dfrac{x}{4} - 3$

_____

_____

_____

So x is _____

**11** $12 < 2x + 4 < 28$

_____

_____

_____

So x is _____

**12** $\dfrac{3 + 2x}{4} + \dfrac{x - 1}{3} < x - 1$

_____

_____

_____

So x is _____

**13** Find the smallest integer that satisfies the inequality $11 < 3(x + 2)$.

_____

_____

_____

_____

**14** Find the smallest integer that satisfies the inequality $15 - 4y < 1 - y$.

_____

_____

_____

_____

## Forming and solving linear inequations

Answer the following, showing your reasoning.

**1**  One plus triple a number is fewer than five times the number minus eight. If the number is a multiple of three, what is the smallest value it can take?

_____

_____

**2**  Three times the sum of double a number and seven is greater than nine fewer than eleven times the number. Write a mathematical statement for the range of values that $x$ can take.

_____

_____

**3**  The area of rectangle A is at least as large as the area of rectangle B. Given that $x > \frac{1}{3}$ (otherwise $3x - 1$ is negative), find the maximum value that $x$ can take.

_____

_____

**4**  The area of triangle A is smaller than the area of triangle B. If $x$ must be an integer, find the maximum value that $x$ can take.

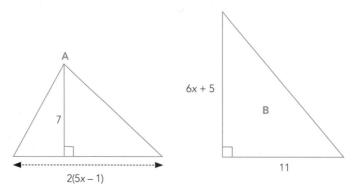

_____

_____

**5**  One of the parallel sides of a trapezium is twice as long as the other ($a$). The vertical height of the trapezium is 5 cm and its area must be 29 cm² at most. If $a$ must be an integer, find the maximum value that $x$ can take.

_____

_____

PHOTOCOPYING OF THIS PAGE IS RESTRICTED UNDER LAW.    ISBN: 9780170477680

# Simultaneous equations

- 'Simultaneous' means 'at the same time'.
- For simultaneous equations, you need to be able to solve two equations that are both true 'at the same time'.
- Because there are two equations, there are also two variables.
- There are two methods for solving these, and which one you use depends on how the equations are structured.

## Substitution

- Substitution is easiest where one of the equations is expressed as $x = ...$ or $y = ....$
- As the name suggests, you substitute the $x = ...$ into the other equation.
- You should number each equation, and say what you are doing at each step.

**Examples:**

**1** Solve the equations $x = 3y + 1$ and $2x + y = 16$.

$$x = 3y + 1 \quad ①$$
$$wa \ 2x + y = 16 \quad ②$$

| | | Number the equations. |

Substitute ① into ②: $\quad 2(3y + 1) + y = 16$

From equation ①, we know that $x$ and $3y + 1$ are equal.

$$6y + 2 + y = 16$$
$$7y + 2 = 16$$
$$7y = 14$$
$$y = 2$$

Substitute for y in ①: $\quad x = 3(2) + 1$
$$x = 7$$
$$(7, 2)$$

Check your answer by substituting into the original equations
① $\quad 7 = 3(2) + 1 \quad ✓$
② $\quad 2(7) + 2 = 16 \quad ✓$

**2** Solve the equations $4y - x = 16$ and $y = 2x - 3$.

$$4y - x = 16 \quad ①$$
$$y = 2x - 3 \quad ②$$

Select whichever equation is easier for substitution.

Substitute ② into ①: $\quad 4(2x - 3) - x = 16$

Select ② because it has the form $y = ....$

$$8x - 12 - x = 16$$
$$7x - 12 = 16$$
$$7x = 28$$
$$x = 4$$

Substitute for x in ②: $\quad y = 2 \times 4 - 3$
$$y = 5$$
$$(4, 5)$$

Again substitute these values into the original equation to check they are correct.

Solve the following simultaneous equations, showing your reasoning.

**1**
$$x = y - 1$$
$$2x + y = 16$$

**2**
$$y = 2x + 4$$
$$3y + x = 26$$

ISBN: 9780170477680 PHOTOCOPYING OF THIS PAGE IS RESTRICTED UNDER LAW.

**3**

$$x = 2y - 5$$
$$3x - y = 25$$

_____

_____

_____

_____

_____

**4**

$$5x + y = 30$$
$$y = 2x + 9$$

_____

_____

_____

_____

_____

**5**

$$y = 4x + 2$$
$$3y - x = 50$$

_____

_____

_____

_____

**6**

$$y = 6x - 6$$
$$4x + y = 14$$

_____

_____

_____

_____

**7**

$$x = 3 - y$$
$$2x - 5y = 20$$

_____

_____

_____

_____

**8**

$$x = 4y$$
$$154 + 2x = y$$

_____

_____

_____

_____

**9**

$$x = 2 - 3y$$
$$4x - 2y - 22 = 0$$

_____

_____

_____

_____

**10**

$$3x - 2y + 34 = 0$$
$$y = 7x + 39$$

_____

_____

_____

_____

PHOTOCOPYING OF THIS PAGE IS RESTRICTED UNDER LAW.    ISBN: 9780170477680

## Elimination

- Elimination is easiest when the two equations have the same structure.

For example:

$$5x + y = 17$$
$$2x + 3y = 12$$

x terms — y terms — = signs — constants

- You solve these by multiplying one or both equations by a constant in order to make either the $x$ terms or the $y$ terms the **same size** but with **different signs**.
- Then **add** the two equations to produce an equation with one variable only.
- You should number each equation, and say what you are doing at each step.

**Examples:**

**1** Solve the equations $3x + y = 12$ and $5x + 4y = 27$.

$$3x + y = 12 \quad ①$$
$$5x + 4y = 27 \quad ②$$
Multiply ① by –4: $\quad -12x - 4y = -48 \quad ③$

> Call this ③ because it is a new equation.

Add ② and ③: $\quad -7x = -21$
$$\therefore \quad x = 3$$
Substitute for x in ①: $\quad 3(3) + y = 12$
$$\therefore \quad y = 3$$
$$(3, 3)$$

> Choose –4 because the $y$ term in ② is +4y.

**2** Solve the equations $6x - 3y = 12$ and $5x - 4y = 1$.

$$6x - 3y = 12 \quad ①$$
$$5x - 4y = 1 \quad ②$$

Multiply ① by –4: $\quad -24x + 12y = -48 \quad ③$
Multiply ② by 3: $\quad 15x - 12y = 3 \quad ④$

> You need to multiply each equation by a different constant to create +12y and –12y.

Add ③ and ④: $\quad -9x = -45$
$$\therefore \quad x = 5$$
Substitute for x in ①: $\quad 6(5) - 3y = 12$
$$-3y = 12 - 30$$
$$\therefore \quad y = 6$$
$$(5, 6)$$

> You should substitute these values into the original equation to check they are correct.

Solve the following simultaneous equations, showing your reasoning.

**1**
$$5x + y = 14$$
$$3x - y = 2$$

**2**
$$7x + 2y = 15$$
$$3x - y = -1$$

_____

_____

_____

_____

_____

ISBN: 9780170477680 PHOTOCOPYING OF THIS PAGE IS RESTRICTED UNDER LAW.

**3**
$$4x - 2y = 24$$
$$5x + y = 16$$

**4**
$$6x + y = 2$$
$$3x + 4y = 8$$

**5**
$$3y - 2x = 15$$
$$y - 4x = 25$$

**6**
$$3x + 4y = 14$$
$$-x + 5y = 8$$

**7**
$$7x + 5y = 20$$
$$2x + 3y = 23$$

**8**
$$3x - 2y = 18$$
$$5x + 4y = 41$$

**9**
$$5x + 3y = 1$$
$$7x - 2y = -11$$

**10**
$$9x + 4y = -12$$
$$2x + 3y = 10$$

PHOTOCOPYING OF THIS PAGE IS RESTRICTED UNDER LAW.
ISBN: 9780170477680

## Forming and solving simultaneous equations

**Hints:**

- Create two equations with two unknown variables.
- Decide whether to use elimination or substitution.
- Make sure you answer the original question. This might involve doing further calculations with your values for $x$ and $y$.

**Example:**

Calculate the values of $x$ and $y$, then state the dimensions of this rectangle.

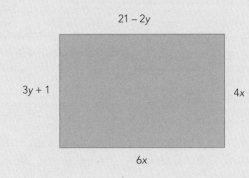

$$6x = 21 - 2y \quad ①$$
$$4x = 3y + 1 \quad ②$$

$$6x + 2y = 21 \quad ③$$
$$4x - 3y = 1 \quad ④$$

Multiply ③ by 1.5: $\quad 9x + 3y = 31.5 \quad ⑤$
Add ⑤ and ④: $\qquad\qquad 13x = 32.5$
$$x = 2.5$$

Substitute for $x$ in ③:
$$6 \times 2.5 + 2y = 21$$
$$2y = 6$$
$$y = 3$$

Base = 6 x 2.5 = 15
Height = 4 x 2.5 = 10

Answer the following, showing your reasoning.

**1** The sum of two numbers is 23.
The difference of the two numbers is 9.
Find the larger number.

**2** The perimeter of a rectangle is 150 cm.
Its base is 15 cm longer than its height.
Calculate the dimensions of the rectangle.

_____

_____

_____

_____

_____

_____

**3**   The difference between the two smaller angles of a right-angled triangle, $x$ and $y$, is 26°. Calculate the values of $x$ and $y$.

_____

_____

_____

_____

_____

_____

**4**   A straight line has a gradient of –1 and intercepts the $y$-axis at 5. A second line has the equation $y = 0.5x + 2$. Where do these two lines intersect?

_____

_____

_____

_____

_____

_____

**5**   Triple the sum of two integers is 120. Half of their difference is 30. Find the two values.

_____

_____

_____

_____

_____

_____

**6**   Two angles in a rectangle are $2x + 3y$ and $5x - 6y$. Calculate the values of $x$ and $y$.

_____

_____

_____

_____

_____

_____

**7**   The perimeter of rectangle X is 34 cm, and the perimeter of rectangle Y is 38 cm. Find the length and the width of each rectangle.

_____

_____

_____

_____

_____

_____

PHOTOCOPYING OF THIS PAGE IS RESTRICTED UNDER LAW.   ISBN: 9780170477680

 Challenge 1

Answer the following.

**1**   Solve the inequality $2(x - 3) - (4x - 2) < 7(2 + x)$.

_____

_____

_____

**2**   Solve $\dfrac{5x - 2}{4} + \dfrac{2x + 3}{5} = 10$.

_____

_____

_____

**3**   Find the value of $x$ when $y = 90°$ and $z = -2$.

$5x - z^2 = \sqrt{\sin y}$

_____

_____

_____

**4**   Solve $2^{x + 4} < 8^x$.

_____

_____

**5**   If $7x - 3y = 29$ and $4x - 7y = 6$, what is the value of $x - y$?

_____

_____

_____

**6**   Solve these simultaneous equations:      $5^{2x} \times 5^y = 125$
$2^{3x} \times 2^{2y} = 128$

_____

_____

_____

ISBN: 9780170477680     PHOTOCOPYING OF THIS PAGE IS RESTRICTED UNDER LAW.

# Quadratic expressions

- An expression in which the highest power of the variable is **2** is known as a **quadratic** expression.
- Usually, these contain an $x^2$, but in factorised form, quadratic expressions may look like $(x \pm a)(x \pm b)$ or $x(x \pm a)$.
- When graphed, quadratics form curves known as **parabolas**:

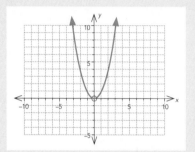   or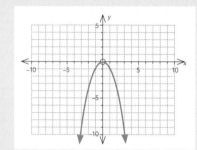

## Expanding quadratic expressions

Remember **FOIL**: multiply the **F**irsts
  **O**uters
  **I**nners
  **L**asts.

| F | O | I | L |

e.g.  $(x + 3)(2x - 7) = 2x^2 - 7x + 6x - 21$

$= 2x^2 - x - 21$

Combine like terms.

Answers are usually written in this order: $x^2$ **term**, **x term**, **number**.

### Examples:

**1**  $(x + 6)(x + 3) = x^2 + 3x + 6x + 18$
  $= x^2 + 9x + 18$

**2**  $(4x + 2)(x - 3) = 4x^2 - 12x + 2x - 6$
  $= 4x^2 - 10x - 6$

**3**  $(x + 5)^2 = (x + 5)(x + 5)$
  $= x^2 + 5x + 5x + 25$
  $= x^2 + 10x + 25$

**4**  $(5x - 3)^2 = (5x - 3)(5x - 3)$
  $= 25x^2 - 15x - 15x + 9$
  $= 25x^2 - 30x + 9$

**5**  $(x + 6)(x - 6) = x^2 - 6x + 6x - 36$
  $= x^2 - 36$

**6**  $(x - 4)(7 - x) = 7x - x^2 - 28 + 4x$
  $= -x^2 + 11x - 28$

Expand and simplify the following.

**1**  $(x + 2)(x + 9)$

**2**  $(x + 5)(x - 4)$

PHOTOCOPYING OF THIS PAGE IS RESTRICTED UNDER LAW.   ISBN: 9780170477680

**3**  $(x + 3)(x - 7)$

_____

_____

**4**  $(x - 4)(x - 8)$

_____

_____

**5**  $(2x + 3)(x - 4)$

_____

_____

**6**  $(3 + x)(x + 9)$

_____

_____

**7**  $(x + 9)^2$

_____

_____

**8**  $(x - 7)^2$

_____

_____

**9**  $(x - 8)(2 + 3x)$

_____

_____

**10**  $(4x + 2)(3x - 5)$

_____

_____

**11**  $(2x - 7)^2$

_____

_____

**12**  $(x + 8)(x - 8)$

_____

_____

**13**  $(6x - 3)(3x - 4)$

_____

_____

**14**  $(4 - x)^2$

_____

_____

**15**  $(2 + 3x)(1 - 8x)$

_____

_____

**16**  $(3x - 1)(3x - 3)$

_____

_____

**17**  $(5 - 3x)(2 - 4x)$

_____

_____

**18**  $(9 - 3x)(9 + 3x)$

_____

_____

## Factorising quadratic expressions

### 1 Where the coefficient of $x^2$ is 1

The sign of the constant tells you the signs in the factors:
$+ \Rightarrow (x - ?)(x - ?)$
or $(x + ?)(x + ?)$
$- \Rightarrow (x + ?)(x - ?)$

**Steps:**

1 List all the factors of the constant: $x^2 + 5x \ominus 24$

1, 24
2, 12
3, 8
4, 6

2 Select the pair that could add or subtract to give the coefficient of x: $x^2 + 5x - 24$

$-3 + 8 = +5$

3 The factors are $(x - 3)(x + 8)$

4 **Check** your answer by expanding the brackets using **FOIL** — you should get the original expression: $(x - 3)(x + 8) = x^2 + 5x - 24$

**Examples:**

1 $x^2 + 9x + 18 = (x + 3)(x + 6)$

1, 18
2, 9
3, 6

2 $x^2 - 12x + 27 = (x - 3)(x - 9)$

1, 27
−1, −27
3, 9
−3, −9

3 $x^2 - 4x - 12 = (x + 2)(x - 6)$

1, −12
−1, 12
2, −6
−2, 6
3, −4
−3, 4

4 $x^2 - 49 = x^2 + 0x - 49 = (x + 7)(x - 7)$

Insert a 'fake' x term.

1, −49
−1, 49
7, −7

Factorise the following.

1 $x^2 + 8x + 7$

_____

_____

2 $x^2 + 7x + 6$

_____

_____

PHOTOCOPYING OF THIS PAGE IS RESTRICTED UNDER LAW.    ISBN: 9780170477680

**3** $x^2 + 12x + 27$

_____

_____

**4** $x^2 + 6x - 16$

_____

_____

**5** $x^2 + 19x + 70$

_____

_____

**6** $x^2 - 9x + 20$

_____

_____

**7** $x^2 - 9x - 22$

_____

_____

**8** $x^2 - 6x + 9$

_____

_____

**9** $x^2 + 12x + 36$

_____

_____

**10** $x^2 - x - 42$

_____

_____

**11** $x^2 - 8x + 16$

_____

_____

**12** $x^2 + 8x - 9$

_____

_____

**13** $x^2 - 81$

_____

_____

**14** $25 - x^2$

_____

_____

**15** $9 - 4x^2$

_____

_____

**16** $1 - 25x^2$

_____

_____

**17** $27 + 12x + x^2$

_____

_____

**18** $21 - 10x + x^2$

_____

_____

### 2 Where the coefficient of $x^2$ is not 1 but there is a common factor

**Steps:**

> **2** divides into the coefficients and the constant.

1 Take out the common factor first:

$2x^2 - 6x - 8 = 2(x^2 - 3x - 4)$
$\phantom{2x^2 - 6x - 8} = 2(x + 1)(x - 4)$

2 Factorise the contents of the bracket:

**Examples:**

1 $\quad 4x^2 + 4x - 24 = 4(x^2 + x - 6)$
$\phantom{4x^2 + 4x - 24} = 4(x + 3)(x - 2)$

2 $\quad 3x^2 + 36x + 108 = 3(x^2 + 12x + 36)$
$\phantom{3x^2 + 36x + 108} = 3(x + 6)(x + 6)$
$\phantom{3x^2 + 36x + 108} = 3(x + 6)^2$

3 $\quad 2x^2 - 32 = 2(x^2 - 16)$
$\phantom{2x^2 - 32} = 2(x - 4)(x + 4)$

Factorise the following.

**19** $2x^2 + 16x + 14$

**20** $5x^2 + 20x + 15$

**21** $3x^2 + 9x - 12$

**22** $4x^2 - 52x + 160$

**23** $6x^2 + 24x + 24$

**24** $10x^2 - 90$

**25** $-2x^2 + 28x - 98$

**26** $3x^2 - 3$

**27** $24 - 4x + 4x^2$

**28** $80 - 5x^2$

PHOTOCOPYING OF THIS PAGE IS RESTRICTED UNDER LAW.     ISBN: 9780170477680

## 3 Where the coefficient of $x^2$ is not 1 and there is no common factor

**Example:**   Factorise $2x^2 + 3x - 20$

Step 1:   Multiply the coefficient of $x^2$ by the constant.

$$2 \times -20 = -40$$

Step 2:   List factors of $-40$:      $-1, 40$      $1, -40$
$-2, 20$      $2, -20$
$-4, 10$      $4, -10$
$-5, 8$       $5, -8$

Step 3:   Select the pair that can add to produce the coefficient of $x$.

$$-5 + 8 = 3$$

Step 4:   Rewrite the quadratic with this pair as coefficients of **two** $x$ terms.

$$2x^2 - 5x + 8x - 20$$

> Note: You can swap the order of $-5x$ and $+8x$ around but **not** the signs.

Step 5:   Factorise the first pair of terms and the second pair of terms.

$$x(2x - 5) + 4(2x - 5)$$

> The contents of both brackets should be the same!

Step 6:   Factorise, making the bracket contents the common factor.

$$(2x - 5)(x + 4) \text{ or } (x + 4)(2x - 5)$$

### Examples showing what you should get at each step:

**1**  $12x^2 - 5x - 3$

Step 1:   $12 \times -3 = -36$

Step 2:   $1, 36$
$2, 18$
$3, 12$
$4, 9$
$6, 6$

> Remember, one of each pair will be a negative.

Step 3:   $+4 - 9 = -5$

Step 4:   $12x^2 + 4x - 9x - 3$

Step 5:   $4x(3x + 1) - 3(3x + 1)$

Step 6:   $(3x + 1)(4x - 3)$

**2**  $10x^2 - 11x + 3$

Step 1:   $10 \times 3 = 30$

Step 2:   $1, 30$
$2, 15$
$3, 10$
$5, 6$

> Factors must be either both positive or both negative.

Step 3:   $-5 - 6 = -11$

Step 4:   $10x^2 - 5x - 6x + 3$

Step 5:   $5x(2x - 1) - 3(2x - 1)$

Step 6:   $(2x - 1)(5x - 3)$

Factorise the following.

**29**   $2x^2 + 11x + 5$

_____

_____

_____

**30**   $2x^2 + 7x + 6$

_____

_____

_____

ISBN: 9780170477680    PHOTOCOPYING OF THIS PAGE IS RESTRICTED UNDER LAW.

**31**  $2x^2 - 7x + 5$

_____

_____

_____

_____

**32**  $2x^2 - x - 10$

_____

_____

_____

_____

**33**  $3x^2 - 13x + 4$

_____

_____

_____

_____

**34**  $8x^2 + 10x - 3$

_____

_____

_____

_____

**35**  $4x^2 - 33x - 27$

_____

_____

_____

_____

**36**  $6x^2 + x - 15$

_____

_____

_____

_____

**37**  $12x^2 + x - 6$

_____

_____

_____

_____

**38**  $9x^2 - 18x + 5$

_____

_____

_____

_____

**39**  $4x^2 - 4x + 1$

_____

_____

_____

_____

**40**  $18x^2 - 2$

_____

_____

_____

_____

PHOTOCOPYING OF THIS PAGE IS RESTRICTED UNDER LAW.    ISBN: 9780170477680

## Simplifying quadratic fractions

Factorise the top and bottom, then cancel the common factors.

**Hint:** If no brackets will cancel out, check that you have factorised the terms correctly.

**Examples:**

1  $\dfrac{x^2 + 9x + 8}{x + 8} = \dfrac{(x + 1)(\cancel{x + 8})}{\cancel{x + 8}}$

   $= (x + 1)$

2  $\dfrac{x^2 + 6x + 9}{x^2 - x - 12} = \dfrac{\cancel{(x + 3)}(x + 3)}{\cancel{(x + 3)}(x - 4)}$

   $= \dfrac{x + 3}{x - 4}$

3  $\dfrac{21 - 6x}{6x^2 + 3x - 84} = \dfrac{3(7 - 2x)}{3(2x^2 + x - 28)}$

   $= \dfrac{-3(2x - 7)}{3(2x - 7)(x + 4)}$

   $= \dfrac{-1}{x + 4}$

4  $\dfrac{x}{x^2 - 25} \div \dfrac{5x}{x + 5} = \dfrac{x}{(x - 5)(x + 5)} \times \dfrac{x + 5}{5x}$

   $= \dfrac{1}{5x - 25}$

   or $\dfrac{1}{5(x - 5)}$

Simplfy these.

1  $\dfrac{2x + 10}{(x + 5)^2}$

_____

_____

2  $\dfrac{(x - 3)^2}{3x - 9}$

_____

_____

3  $\dfrac{x^2 + 5x}{x + 5}$

_____

_____

4  $\dfrac{4x^2 - 16}{x - 2}$

_____

_____

5  $\dfrac{x^2 + 11x + 30}{x + 6}$

_____

_____

6  $\dfrac{x^2 - 4x - 21}{x - 7}$

_____

_____

7  $\dfrac{x + 6}{x^2 + 5x - 6}$

_____

_____

8  $\dfrac{5x + 15}{x^2 - x - 12}$

_____

_____

ISBN: 9780170477680 PHOTOCOPYING OF THIS PAGE IS RESTRICTED UNDER LAW.

**9**    $\dfrac{x^2 + 11x + 18}{x^2 - 6x - 16}$

**10**    $\dfrac{x^2 - 7x + 6}{x^2 - 13x + 42}$

**11**    $\dfrac{x^2 - 1}{x + 1}$

**12**    $\dfrac{x - 3}{9 - x^2}$

**13**    $\dfrac{49 - x^2}{3x^2 - 16x - 35}$

**14**    $\dfrac{9x^2 + 49x + 20}{25 - x^2}$

**15**    $\dfrac{x^2 - 6x + 9}{5x^2 - 11x - 12}$

**16**    $\dfrac{21x^2 + 22x - 8}{12x^2 + 19x + 4}$

**17**    $\dfrac{x^2 - 1}{3x + 9} \times \dfrac{x + 3}{x^2 - x}$

**18**    $\dfrac{x^2}{2x^2 - x - 15} \div \dfrac{-x}{x - 3}$

**19**    $\dfrac{x^2 + 5x - 6}{x^2 - 25} \div \dfrac{1 - x}{x + 5}$

**20**    $\dfrac{1}{x^2 + 10x + 25} \div \dfrac{2}{x + 5}$

PHOTOCOPYING OF THIS PAGE IS RESTRICTED UNDER LAW.    ISBN: 9780170477680

# Solving quadratic equations

## 1 Factorised quadratic equations

$$\text{If } a \times b = 0$$

then either $\boxed{a = 0}$ or $\boxed{b = 0}$ or $a = 0$ and $b = 0$

Remember, the x (times) sign has not been written in here.

So, if we have a quadratic equation $(x - 1)(x + 4) = 0$

then either $\boxed{(x - 1) = 0}$ or $\boxed{(x + 4) = 0}$

So either $x = 1$ or $x = -4$

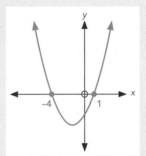

**Note:**
- Most quadratic equations that you come across will have **two solutions**. However, some have just **one solution**, and some have **no real solutions**.
- In practical situations, there are usually two solutions, but only one works.

**Examples:**

**1** $(x + 8)(x - 3) = 0$

Either $(x + 8) = 0$

$x = -8$

or $(x - 3) = 0$

$x = 3$

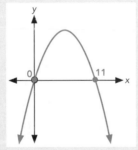

**2** $(x + 5)^2 = 0$

$\therefore (x + 5) = 0$

$x = -5$

$\boxed{\text{One solution only}}$

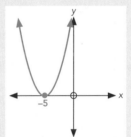

**3** $6x(11 - x) = 0$

Clearly $6 \neq 0$

So either $x = 0$

or $(11 - x) = 0$

$x = 11$

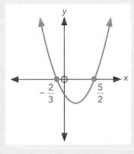

Note: $66x - 6x^2 = 0$

$-6x^2 \Rightarrow$ parabola is upside down.

**4** $(3x + 2)(2x - 5) = 0$

Either $(3x + 2) = 0$

$3x = -2$

$x = -\dfrac{2}{3}$

or $(2x - 5) = 0$

$2x = 5$

$x = \dfrac{5}{2}$

43

Solve the following equations.

**1** $(x + 8)(x - 2) = 0$

_____

_____

_____

**2** $(x - 6)(x + 4) = 0$

_____

_____

_____

ISBN: 9780170477680 PHOTOCOPYING OF THIS PAGE IS RESTRICTED UNDER LAW.

**3**  $(x + 2)(x + 7) = 0$

_____

_____

**4**  $(x - 4)(x - 1) = 0$

_____

_____

**5**  $(x + 5)(x - 9) = 0$

_____

_____

**6**  $(x - 6)^2 = 0$

_____

_____

**7**  $(x + 8)^2 = 0$

_____

_____

**8**  $(x - \dfrac{1}{2})^2 = 0$

_____

_____

**9**  $(x + 3)(2 - x) = 0$

_____

_____

_____

**10**  $(x - 3)(6 + x) = 0$

_____

_____

**11**  $x(x - 2) = 0$

_____

_____

_____

**12**  $3x(x + 4) = 0$

_____

_____

**13**  $-x(4x - 3) = 0$

_____

_____

_____

**14**  $5x(2x + 7) = 0$

_____

_____

**15**  $-9x(x - 4) = 0$

_____

_____

_____

**16**  $\dfrac{2}{3}x(x - \dfrac{3}{4}) = 0$

_____

_____

_____

PHOTOCOPYING OF THIS PAGE IS RESTRICTED UNDER LAW.    ISBN: 9780170477680

## 2 Unfactorised quadratic equations

Solve these equations by factorising the equation first.

**Examples:**

**1** Solve $x^2 - x - 12 = 0$

Step 1: Factorise $(x + 3)(x - 4) = 0$

Step 2: Solve by considering what happens when each bracket equals 0.

Either $(x + 3) = 0$

$x = -3$

or $(x - 4) = 0$

$x = 4$

So, if $x^2 - x - 12 = 0$, then $x$ is either $-3$ or $4$.

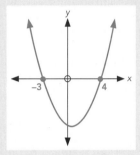

**2**
$x^2 + 6x + 5 = 0$
$(x + 5)(x + 1) = 0$
Either $(x + 5) = 0$
$x = -5$
or $(x + 1) = 0$
$x = -1$

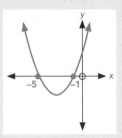

**3**
$x^2 + 12x + 36 = 0$
$(x + 6)(x + 6) = 0$
$\therefore (x + 6) = 0$
$x = -6$

One solution only

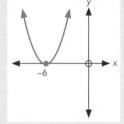

**4**
$x^2 - 81 = 0$
$(x + 9)(x - 9) = 0$
Either $(x + 9) = 0$
$x = -9$
or $(x - 9) = 0$
$x = 9$

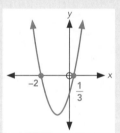

**5**
$6x^2 + 10x - 4 = 0$
$(2x + 4)(3x - 1) = 0$
Either $(2x + 4) = 0$
$2x = -4$
$x = -2$
or $(3x - 1) = 0$
$x = \dfrac{1}{3}$

**6**
$-x^2 - 3x = 0$
$-x(x + 3) = 0$
Either $x = 0$
or $(x + 3) = 0$
$x = -3$

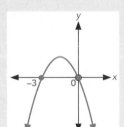

**7**
$16 + x^2 - 8x = 0$
$x^2 - 8x + 16 = 0$
$(x - 4)^2 = 0$
$x = 4$

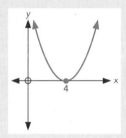

Solve the following equations.

**17** $x^2 + 9x + 8 = 0$

_____

_____

**18** $x^2 + 4x - 12 = 0$

_____

_____

ISBN: 9780170477680    PHOTOCOPYING OF THIS PAGE IS RESTRICTED UNDER LAW.

**19**  $x^2 - 6x + 5 = 0$

**20**  $x^2 - 8x = 0$

**21**  $x^2 + 6x - 16 = 0$

**22**  $x^2 + 10x + 25 = 0$

**23**  $x^2 + 6x = 0$

**24**  $x^2 + 5x - 36 = 0$

**25**  $x^2 - 16x + 64 = 0$

**26**  $x^2 - x - 42 = 0$

**27**  $x^2 + 18x + 81 = 0$

**28**  $x^2 - 9 = 0$

**29**  $x^2 - 6x - 72 = 0$

**30**  $x^2 - 2x - 80 = 0$

**31**  $3x^2 + 13x + 4 = 0$

**32**  $4x^2 - 12x + 5 = 0$

**33**  $2x^2 - x - 28 = 0$

**34**  $12x^2 - 27x + 6 = 0$

PHOTOCOPYING OF THIS PAGE IS RESTRICTED UNDER LAW.    ISBN: 9780170477680

### 3 Quadratic equations where rearrangement is needed

- Sometimes quadratic equations are not organised like the ones in the previous exercise.
- Use the equation-solving rules to reorganise them into an $ax^2 + bx + c = 0$ form.
- Then factorise and solve as usual.

**Examples:**

1
$$x^2 - 10x = -21 \qquad (+\ 21)$$
$$x^2 - 10x + 21 = 0$$
$$(x - 7)(x - 3) = 0$$
$$x = 7 \text{ or } 3$$

2
$$10x + 16 = -x^2 \qquad (+\ x^2)$$
$$x^2 + 10x + 16 = 0$$
$$(x + 2)(x + 8) = 0$$
$$x = -2 \text{ or } -8$$

3
$$3x = x^2 - 18 \qquad (-\ 3x)$$
$$x^2 - 3x - 18 = 0$$
$$(x + 3)(x - 6) = 0$$
$$x = -3 \text{ or } 6$$

4
$$x^2 - 8x + 3 = 12 \qquad (-\ 12)$$
$$x^2 - 8x - 9 = 0$$
$$(x + 1)(x - 9) = 0$$
$$x = -1 \text{ or } 9$$

5
$$(x - 7)(x + 4) = -18$$
$$x^2 - 3x - 28 = -18 \qquad (+\ 18)$$
$$x^2 - 3x - 10 = 0$$
$$(x + 2)(x - 5) = 0$$
$$x = -2 \text{ or } 5$$

6
$$\frac{40 - 6x}{x} = x \qquad (\times\ x)$$
$$40 - 6x = x^2 \qquad (-\ 40,\ +\ 6x)$$
$$x^2 + 6x - 40 = 0$$
$$(x + 10)(x - 4) = 0$$
$$x = -10 \text{ or } 4$$

Solve the following equations.

**35** $x^2 - 5x = 24$

**36** $x^2 - 70 = 3x$

**37** $27 - 6x = x^2$

**38** $18 = 9x - x^2$

**39** $x(x - 7) = 44$

**40** $x = \dfrac{24 - 5x}{x}$

**41** $(x - 2)(x - 9) = 60$

**42** $2x^2 + 30 = x(x - 11)$

ISBN: 9780170477680    PHOTOCOPYING OF THIS PAGE IS RESTRICTED UNDER LAW.

## Forming quadratic equations

- Substitute the information you have been given into the appropriate area or volume formula.
- Expand any brackets.
- Then simplify by collecting like terms.

### 1 Equations involving lengths and areas

**Example:** The area of triangle A is greater than the area of triangle B. Find an expression for the difference in their areas.

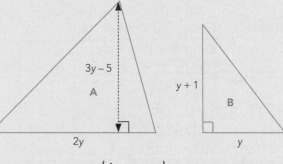

$$\frac{1}{2}(2y \times (3y-5)) - \frac{1}{2}(y \times (y+1)) = (3y^2 - 5y) - \left(\frac{1}{2}(y^2 + y)\right)$$

$$= 3y^2 - 5y - \frac{1}{2}y^2 - \frac{1}{2}y$$

$$= 2.5y^2 - 5.5y$$

Answer the following questions. Show your reasoning.

**1** Write an expression for the difference between the areas of these squares.

_____
_____
_____
_____

**2** Show that the shaded area is $15z^2 - 9z + 1$.

_____
_____
_____
_____

**3** The area of rectangle A is twice the area of rectangle B.
Show that $3x - 5 = 0$.

_____
_____
_____
_____

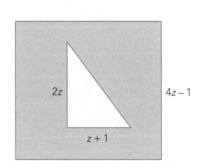
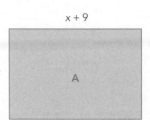
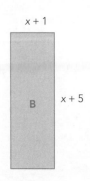

PHOTOCOPYING OF THIS PAGE IS RESTRICTED UNDER LAW.   ISBN: 9780170477680

**4** Write an expression for the area of this shape.

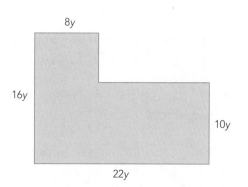

_____

_____

_____

_____

**5** The area of this trapezium is 38 cm².
Show that $5x^2 + 26x - 71 = 0$.

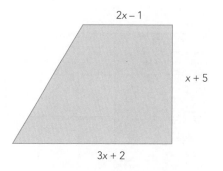

_____

_____

_____

_____

**6** The radius of the larger circle is $2z + 1$.
The radius of the smaller circle is $z - 1$.
Show that the shaded area is $3\pi z(z + 2)$.

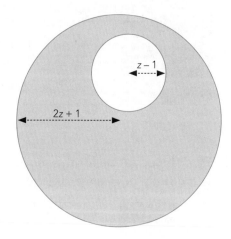

_____

_____

_____

_____

**7** Write an expression in terms of $x$ and $\pi$ for the shaded
area. What fraction of the semicircle is shaded?

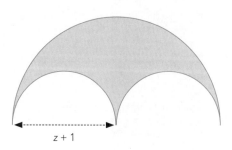

_____

_____

_____

_____

**8** A cylinder with a radius of $a + 3$ has a height of twice its radius. Write an expression in terms of
$a$ and $\pi$ for its entire surface area.

_____

_____

_____

## 2 Equations from number patterns

- You know how to find an equation of a **linear** sequence by looking at **differences** between terms.
- You can use this knowledge, along with a little more to find equations of quadratic sequences.

**If the second difference between terms = 2, then the equation takes the form $y = x^2 \pm \ldots$**

### Examples:

**1** Find the equation for the sequence $-1, 2, 7, 14, 23, \ldots$
Look at the differences between each term:

| $x$ | 1 | | 2 | | 3 | | 4 | | 5 |
|---|---|---|---|---|---|---|---|---|---|
| $y$ | −1 | | 2 | | 7 | | 14 | | 23 |
| First difference | | 3 | | 5 | | 7 | | 9 | |
| Second difference | | | 2 | | 2 | | 2 | | |
| $x^2$ | 1 | | 4 | | 9 | | 16 | | 25 |
| $y - x^2$ | −2 | | −2 | | −2 | | −2 | | −2 |

Unlike linear sequences, the first differences are **not** constant.

Second differences **are** constant.
Second difference = 2 ⇒
Equation is $y = x^2 \pm \ldots$

Calculate $x^2$.

The remainder is constant
⇒ $y - x^2 = -2$, so $y = x^2 - 2$

**2** Find the equation for the sequence $4, 5, 8, 13, 20, \ldots$
Look at the differences between each term:

| $x$ | 1 | | 2 | | 3 | | 4 | | 5 |
|---|---|---|---|---|---|---|---|---|---|
| $y$ | 4 | | 5 | | 8 | | 13 | | 20 |
| First difference | | 1 | | 3 | | 5 | | 7 | |
| Second difference | | | 2 | | 2 | | 2 | | |
| $x^2$ | 1 | | 4 | | 9 | | 16 | | 25 |
| $y - x^2$ | 3 | | 1 | | −1 | | −3 | | −5 |

Second differences **are** constant.
Second difference = 2.

Calculate $x^2$.

The remainder is linear. Calculate its equation as you would for a straight line:
⇒ $y - x^2 = -2x + 5$
so $y = x^2 - 2x + 5$

PHOTOCOPYING OF THIS PAGE IS RESTRICTED UNDER LAW.
ISBN: 9780170477680

Complete the tables to find the equation for each sequence.

**9**  4, 7, 12, 19, 28, …

| $x$ | 1 | | 2 | | 3 | | 4 | | 5 |
|---|---|---|---|---|---|---|---|---|---|
| $y$ | 4 | | 7 | | 12 | | 19 | | 28 |
| First difference | | | | | | | | | |
| Second difference | | | | | | | | | |
| $x^2$ | | | | | | | | | |
| $y - x^2$ | | | | | | | | | |

_____

Equation: _____

**10**  3, 8, 15, 24, 35, …

| $x$ | 1 | | 2 | | 3 | | 4 | | 5 |
|---|---|---|---|---|---|---|---|---|---|
| $y$ | | | | | | | | | |
| First difference | | | | | | | | | |
| Second difference | | | | | | | | | |
| $x^2$ | | | | | | | | | |
| $y - x^2$ | | | | | | | | | |

_____

Equation: _____

**11**  2, 8, 16, 26, 38, …

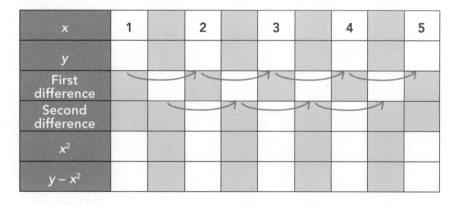

| $x$ | 1 | | 2 | | 3 | | 4 | | 5 |
|---|---|---|---|---|---|---|---|---|---|
| $y$ | | | | | | | | | |
| First difference | | | | | | | | | |
| Second difference | | | | | | | | | |
| $x^2$ | | | | | | | | | |
| $y - x^2$ | | | | | | | | | |

_____

Equation: _____

ISBN: 9780170477680    PHOTOCOPYING OF THIS PAGE IS RESTRICTED UNDER LAW.

**Where** $y = ax^2 \pm \dots$
- Second differences are not always 2.
- They may be **negative**, in which case $y = -ax^2 \pm \dots$
- If they are **4**, then $y = 2x^2 \pm \dots$
  **6**, then $y = 3x^2 \pm \dots$
  **8**, then $y = 4x^2 \pm \dots$
  **1**, then $y = \frac{1}{2}x^2 \pm \dots$, etc.

**Examples:**

**1** Find the equation for the sequence 6, 17, 36, 63, 98, …

| $x$ | 1 | | 2 | | 3 | | 4 | | 5 |
|---|---|---|---|---|---|---|---|---|---|
| $y$ | 6 | | 17 | | 36 | | 63 | | 98 |
| First difference | | 11 | | 19 | | 27 | | 35 | |
| Second difference | | | 8 | | 8 | | 8 | | |
| $4x^2$ | 4 | | 16 | | 36 | | 64 | | 100 |
| $y - 4x^2$ | 2 | | 1 | | 0 | | –1 | | –2 |

Second differences **are** constant.
Second difference = **8**.

Calculate $4x^2$.

The remainder is $-1x + 3$
$\Rightarrow y - 4x^2 = -1x + 3$
so $y = 4x^2 - x + 3$

**2** Find the equation for the sequence –2, –9, –22, –41, –66, …

| $x$ | 1 | | 2 | | 3 | | 4 | | 5 |
|---|---|---|---|---|---|---|---|---|---|
| $y$ | –2 | | –9 | | –22 | | –41 | | –66 |
| First difference | | –7 | | –13 | | –19 | | –25 | |
| Second difference | | | –6 | | –6 | | –6 | | |
| $-3x^2$ | –3 | | –12 | | –27 | | –48 | | –75 |
| $y - (-3x^2)$ $= y + 3x^2$ | 1 | | 3 | | 5 | | 7 | | 9 |

Second differences **are** constant.
Second difference = **–6**.

Calculate $-3x^2$.

The remainder is $2x - 1$
$\Rightarrow y + 3x^2 = 2x - 1$
so $y = -3x^2 + 2x - 1$

PHOTOCOPYING OF THIS PAGE IS RESTRICTED UNDER LAW.  ISBN: 9780170477680

Complete the tables to find the equation for each sequence.

**12**   7, 24, 51, 88, 135, …

| x | 1 | | 2 | | 3 | | 4 | | 5 |
|---|---|---|---|---|---|---|---|---|---|
| y | 7 | | 24 | | 51 | | 88 | | 135 |
| First difference | | | | | | | | | |
| Second difference | | | | | | | | | |
| ___ $x^2$ | | | | | | | | | |
| y − ___ $x^2$ | | | | | | | | | |

Equation: _____

**13**   −2, −1, 4, 13, 26, …

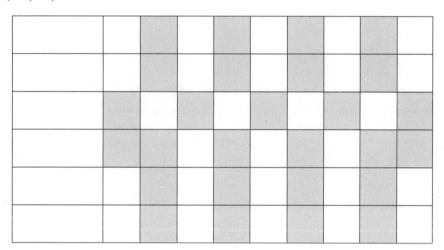

Equation: _____

**14**   2, 6, 8, 8, 6, …

Equation: _____

ISBN: 9780170477680    PHOTOCOPYING OF THIS PAGE IS RESTRICTED UNDER LAW.

## 3 Equations involving groups of numbers

- Read what the question asks for. Call the unknown number $x$ or any other letter.
- If there are two things to find, call the smaller one $x$ and the bigger one ($x \pm ?$).

### Useful expressions:
- Even numbers: call these $2x$, e.g. 10, where $x = 5$.
- Odd numbers: call these $2x + 1$, e.g. 15, where $x = 7$.
- Consecutive numbers come after each other: call these $x$, $x + 1$, $x + 2$, e.g. 6, 7, 8, where $x = 6$.
- Consecutive even numbers: call these $2x$, $2x + 2$, $2x + 4$, e.g. 6, 8, 10, where $x = 3$.
- Consecutive odd numbers: call these $2x + 1$, $2x + 3$, $2x + 5$, e.g. 9, 11, 13, where $x = 4$.
- Alternate means every second item, e.g. 5 and 9 are alternate odd numbers, 8 and 12 are alternate even numbers.

### Examples:
**1** Write an expression for the sum of the squares of two consecutive numbers.

$$\text{Sum} = x^2 + (x + 1)^2$$
$$= x^2 + x^2 + 2x + 1$$
$$= 2x^2 + 2x + 1$$

**2** Consider the sequence $T_n = n^2 + 2n - 1$. Write an expression for the difference between consecutive terms.

The $(n + 1)$th term – the $n$th term $= [(n + 1)^2 + 2(n + 1) - 1] - [n^2 + 2n - 1]$
$$= [(n^2 + 2n + 1) + 2(n + 1) - 1] - [n^2 + 2n - 1]$$
$$= n^2 + 2n + 1 + 2n + 2 - 1 - n^2 - 2n + 1$$
$$= 2n + 3$$

48

Answer the following questions. Show your reasoning.

**15 a** Given that the smaller number is written as $2x + 1$, show that the product of two consecutive odd numbers can be written as $4x^2 + 8x + 3$.

_____

_____

_____

**b** Show that this expression works if the numbers are 11 and 13.

_____

_____

_____

**16 a** Given that the number is written as $2x + 1$, show that the sum of an odd number and its square can be written as $4x^2 + 6x + 2$.

_____

_____

_____

**b** Show that this expression works if the odd number is 15.

_____

_____

_____

PHOTOCOPYING OF THIS PAGE IS RESTRICTED UNDER LAW.    ISBN: 9780170477680

**17** Given that the smaller number is written as $2x + 1$, write an expression for the product of any two alternate odd numbers.

_____

_____

_____

_____

_____

**18** Given that the smaller number is written as $2x + 1$, write an expression for the square of the sum of two consecutive odd numbers.

_____

_____

_____

_____

_____

_____

**19** Given that the smaller number is written as $2x$, write an expression for the sum of the squares of two consecutive even numbers.

_____

_____

_____

_____

_____

**20** Consider the sequence $T_n = n^2 + 3$. Write and simplify an expression for the difference between any two consecutive terms.

_____

_____

_____

_____

_____

**21** Consider the sequence $T_n = n^2 - 3n + 2$. Write and simplify an expression for the sum of any two consecutive terms.

_____

_____

_____

_____

_____

## Forming and solving quadratic equations

- Usually you will get two answers. **Test both to see if they work.** If you have to reject one answer because it isn't practical, **explain why.**

**Examples:**

**1** A rectangle has sides of $2x + 3$ and $x - 2$ metres long. Its area is 60 m$^2$. Calculate its dimensions.

$$(2x + 3)(x - 2) = 60$$
$$2x^2 - x - 6 = 60$$
$$2x^2 - x - 66 = 0$$
$$(2x + 11)(x - 6) = 0$$
$$x = -5.5 \text{ or } 6$$

Sides $2x + 3$ and $x - 2$ cannot be negative so $x = 6$. ⸺ | Usually only one answer will work. You must explain why.
Rectangle dimensions are $2(6) + 3 = 15$ m
and $6 - 2 = 4$ m

**2** The square of a number is equal to 35 more than twice the number. Find the number.

$$x^2 = 2x + 35$$
$$x^2 - 2x - 35 = 0$$
$$(x - 7)(x + 5) = 0$$
$$x = 7 \text{ or } -5$$

Check answers: $\quad x = 7 \Rightarrow 49 = 14 + 35 \qquad ✓$
$\qquad\qquad\qquad x = -5 \Rightarrow 25 = -10 + 35 \qquad ✓$
$\therefore$ Both answers work, so $x$ can be 7 or $-5$.

**3** $x^2 + ax - 11 = 13$, where $a$ is a negative number. If the difference between the solutions to the equation is 10, find the value of $a$, and the possible solutions to the equation.

Step 1:      Reorganise to give 0 on one side:      $x^2 + ax - 24 = 0$

Step 2:      List the factors of 24:      1, 24
     2, 12 ⸺ | **Both** could give solutions with a
     3, 8
     4, 6 ⸺ difference of 10.

Step 3:      Decide which pair of numbers to use:      Have '$-24$' in Step 1 $\Rightarrow$ solutions must have different signs.
$\therefore$ Use 4 and $-6$, or $-4$ and 6.

Step 4:      Decide which pair of factors to use:      $(x - 4)(x + 6)$ or $(x + 4)(x - 6)$
Because $a$ must be negative, we must use the second pair.

$$(x + 4)(x - 6) = x^2 - 2x - 24$$
$\therefore a = -2$ and solutions to the equation are $x = -4$ or 6.

PHOTOCOPYING OF THIS PAGE IS RESTRICTED UNDER LAW.   ISBN: 9780170477680

Answer the following questions. Show your reasoning.

49

**1**   A rectangle has sides of $x - 2$ and $x + 6$ metres long. Its area is 84 m². Calculate its dimensions.

_____

_____

_____

_____

**2**   A circle has an area of 50 cm². Calculate its diameter.

_____

_____

_____

_____

**3**   Find a positive number that is 132 less than its square.

_____

_____

_____

_____

**4**   The square of an integer plus the square of four less than the integer comes to 10. Find the value of the integer.

_____

_____

_____

_____

**5**   What is the perimeter of a rectangle whose length is 11 cm more than its height and whose area is 42 cm²?

_____

_____

_____

_____

**6**   The perpendicular sides of a triangle are $x + 4$ and $2x - 4$. Its area is 16 m². Calculate the lengths of its perpendicular sides.

_____

_____

_____

_____

**7**   The shaded area in this figure is 51 m². Calculate the value of $x$.

_____

_____

_____

_____

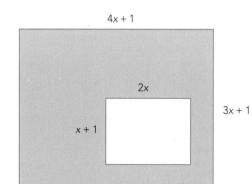

**8** The area of the shaded ring is 62.83 m². Calculate the width of the shaded ring.

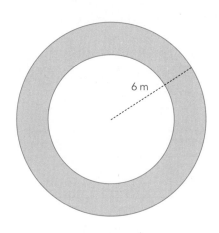

6 m

**9** $x^2 + ax + 15 = 9$, where a is a positive number. If the difference between the solutions to the equation is 1, find the value of a, and the possible solutions to the equation.

**10** $x^2 + ax - 11 = 1$, where a is a negative number. If the difference between the solutions to the equation is 7, find the value of a, and the possible solutions to the equation.

**11** $x^2 + ax - 20 = 5$, where a is a positive number. If the difference between the solutions to the equation is 8, find the value of a, and the possible solutions to the equation.

**12** $x^2 + ax + 40 = 4$, where a is a negative number. If the solutions to the equation are ±p, find the possible values for a, and the solutions.

PHOTOCOPYING OF THIS PAGE IS RESTRICTED UNDER LAW. ISBN: 9780170477680

# Plotting quadratic equations

### Plotting quadratics

- Graphs of quadratic equations form a curve called a parabola.
- A number of programs will plot parabolas for you.
- However, if a hand sketch is required, tables are a useful tool.
- The turning point of a parabola is called the vertex.
- Note the vertex is not pointed.

**Example:**  Draw the graph of $y = x^2$.

| x | y |
|---|---|
| 3 | 9 |
| 2 | 4 |
| 1 | 1 |
| 0 | 0 |
| −1 | 1 |
| −2 | 4 |
| −3 | 9 |

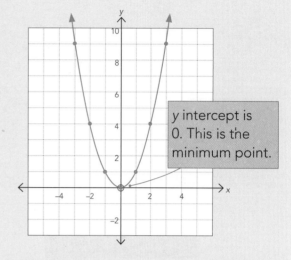

y intercept is 0. This is the minimum point.

- *All* parabolas are this shape, but they can be shifted, turned upside down or stretched, or any combination of these. Here are some examples:

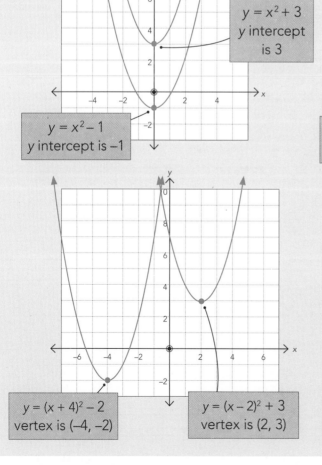

$y = x^2 + 3$
y intercept is 3

$y = x^2 - 1$
y intercept is −1

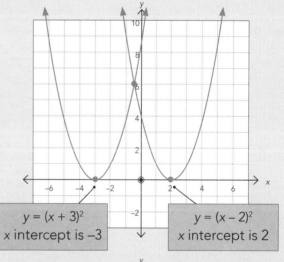

$y = (x + 3)^2$
x intercept is −3

$y = (x - 2)^2$
x intercept is 2

$y = (x + 4)^2 - 2$
vertex is (−4, −2)

$y = (x - 2)^2 + 3$
vertex is (2, 3)

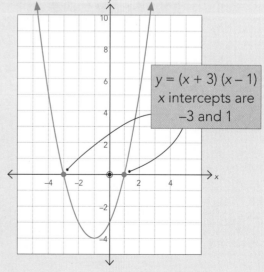

$y = (x + 3)(x - 1)$
x intercepts are −3 and 1

ISBN: 9780170477680    PHOTOCOPYING OF THIS PAGE IS RESTRICTED UNDER LAW.

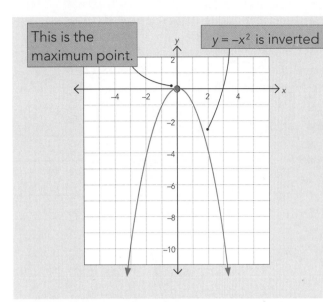

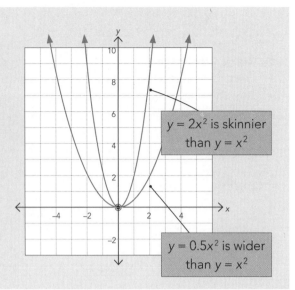

Complete the tables and draw the graphs for the following equations.

**1**  $y = x^2 + 1$

| x | y |
|---|---|
| 3 | |
| 2 | |
| 1 | |
| 0 | |
| −1 | |
| −2 | |
| −3 | |

The vertex is a minimum/maximum

at (_____, _____)

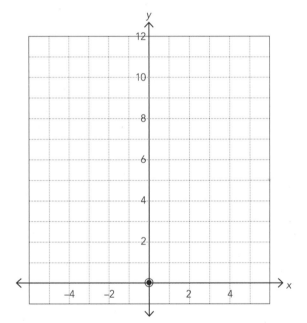

**2**  $y = (x - 2)^2 + 1$

| x | y |
|---|---|
| 5 | |
| 4 | |
| 3 | |
| 2 | |
| 1 | |
| 0 | |
| −1 | |

The vertex is a minimum/maximum

at (_____, _____)

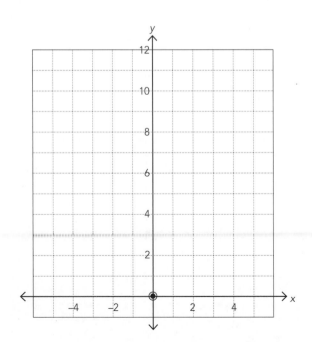

PHOTOCOPYING OF THIS PAGE IS RESTRICTED UNDER LAW.    ISBN: 9780170477680

**3**   $y = -(x + 1)(x - 5)$

| x | y |
|---|---|
| 5 | |
| 4 | |
| 3 | |
| 2 | |
| 1 | |
| 0 | |
| −1 | |

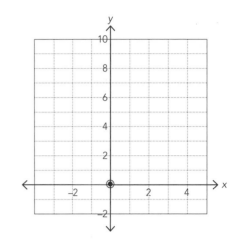

The vertex is a minimum/maximum at (_____, _____)

**4**   $y = 0.75x^2$

| x | y |
|---|---|
| 4 | |
| 2 | |
| 0 | |
| −2 | |
| −4 | |

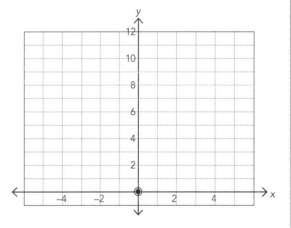

The vertex is a minimum/maximum at (_____, _____)

**5**   $y = 2(x - 1)^2 + 3$

| x | y |
|---|---|
| 3 | |
| 2 | |
| 0 | |
| −1 | |
| −2 | |

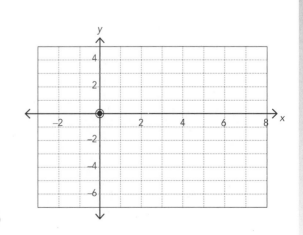

The vertex is a minimum/maximum at (_____, _____)

**6**   $y = -0.5(x - 3)^2 + 4$

| x | y |
|---|---|
| 7 | |
| 5 | |
| 3 | |
| 1 | |
| −1 | |

The vertex is a minimum/maximum at (_____, _____)

# Writing equations for parabolas

- Start by marking points on the graph where x and y are both integers.
- It will pass through *either* the x intercepts

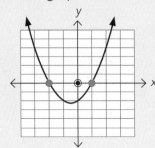

*or* the turning point.

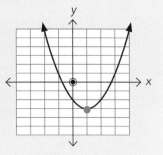

## 1 The x intercepts

- The equation takes the form **y = a(x ± p)(x ± q)**.

**Examples:** Write the equation of the following parabolas.

**1**

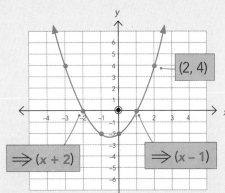

(2, 4)

⟹ (x + 2)  ⟹ (x – 1)

Steps:
1 Mark all the integral points (•).
2 x intercepts ⟹ equation must be
$$y = a(x - 1)(x + 2)$$
3 Substitute the coordinates of another integral point, e.g. (2, 4):
$$4 = a(2 - 1)(2 + 2)$$
$$4 = 4a$$
$$\therefore a = 1$$
So the equation must be **y = (x – 1)(x + 2)**.

**2**

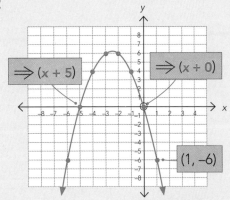

⟹ (x + 5)  ⟹ (x + 0)

(1, –6)

Steps:
1 Mark all the integral points (•).
2 x intercepts ⟹ equation must be
$$y = a(x + 0)(x + 5) = ax(x + 5)$$
3 Substitute the coordinates of another integral point, e.g. (1, –6):
$$-6 = a1(1 + 5)$$
$$-6 = 6a$$
$$\therefore a = -1$$
So the equation must be **y = –x(x + 5)**.

**3**

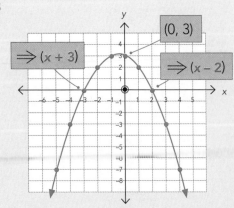

(0, 3)

⟹ (x + 3)  ⟹ (x – 2)

Steps:
1 Mark all the integral points (•).
2 x intercepts ⟹ equation must be
$$y = a(x - 2)(x + 3)$$
3 Substitute the coordinates of another integral point, e.g. (0, 3):
$$3 = a(0 - 2)(0 + 3)$$
$$3 = -6a$$
$$\therefore a = -\frac{1}{2}$$
So the equation must be **$y = -\frac{1}{2}(x - 2)(x + 3)$**.

PHOTOCOPYING OF THIS PAGE IS RESTRICTED UNDER LAW.    ISBN: 9780170477680

Write equations for the following parabolas.

**1**

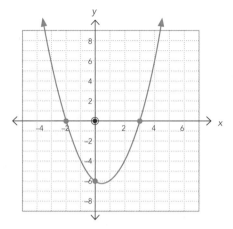

**1** Mark all the integral points (•).

**2** x intercepts ⇒ equation must be

$$y = a(x _____)(x _____)$$

**3** Substitute the coordinates of another

integral point, e.g. (\_\_\_\_, \_\_\_\_)

_____

_____

So the equation must be y = _____

**2**

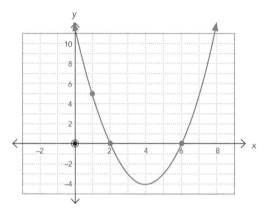

**1** Mark all the integral points (•).

**2** x intercepts ⇒ equation must be

$$y = a(x _____)(x _____)$$

**3** Substitute the coordinates of another

integral point, e.g. (\_\_\_\_, \_\_\_\_)

_____

_____

So the equation must be y = _____

**3**

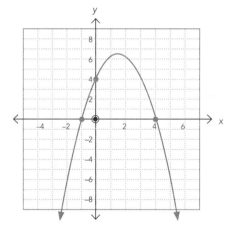

_____

_____

_____

_____

_____

_____

**4**

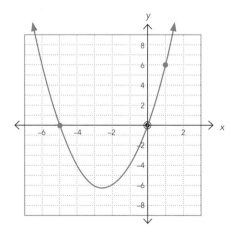

_____

_____

_____

_____

_____

_____

ISBN: 9780170477680     PHOTOCOPYING OF THIS PAGE IS RESTRICTED UNDER LAW.

**5**

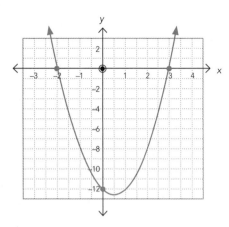

_____

_____

_____

_____

_____

_____

**6**

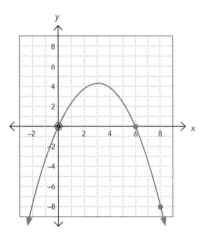

_____

_____

_____

_____

_____

_____

_____

**7**

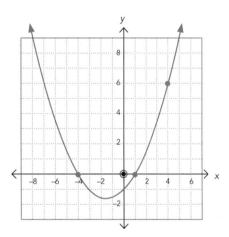

_____

_____

_____

_____

_____

_____

_____

**8**

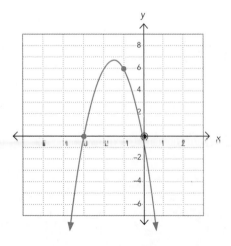

_____

_____

_____

_____

_____

_____

PHOTOCOPYING OF THIS PAGE IS RESTRICTED UNDER LAW.    ISBN: 9780170477680

## 2  The turning point

- The equation takes the form $y = a(x \pm h)^2 \pm v$,
  where a = scale factor, h = horizontal shift, v = vertical shift.

**Examples:** Write the equation of each parabola.

**1**

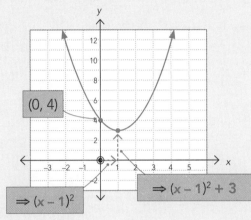

$\Rightarrow (x - 1)^2$

$\Rightarrow (x - 1)^2 + 3$

(0, 4)

Steps:

**1** Mark the turning point and y intercept (•).

**2** Turning point moved **1 unit to right**
$$\Rightarrow y = a(\boldsymbol{x - 1})^2 \pm v$$

**3** Turning point moved **3 units up**
$$\Rightarrow y = a(x - 1)^2 + \boldsymbol{3}$$

**4** Substitute the coordinates of the y intercept, (0, 4):
$$4 = a(0 - 1)^2 + 3$$
$$a = 1$$
So the equation must be $y = (x - 1)^2 + 3$.

**2**

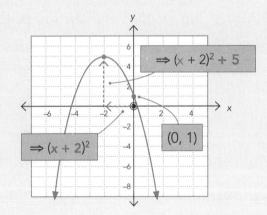

$\Rightarrow (x + 2)^2 + 5$

$\Rightarrow (x + 2)^2$

(0, 1)

Steps:

**1** Mark the turning point and y intercept (•).

**2** Turning point moved **2 units to left**
$$\Rightarrow y = a(\boldsymbol{x + 2})^2 \pm v$$

**3** Turning point moved **5 units up**
$$\Rightarrow y = a(x + 2)^2 + \boldsymbol{5}$$

**4** Substitute the coordinates of the
y intercept, (0, 1):
$$1 = a(0 + 2)^2 + 5$$
$$-4 = 4a$$
$$a = -1$$
So the equation must be $y = -(x + 2)^2 + 5$.

**3**

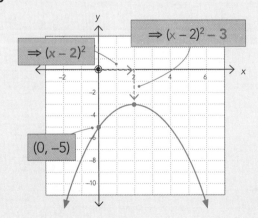

$\Rightarrow (x - 2)^2$

$\Rightarrow (x - 2)^2 - 3$

(0, −5)

Steps:

**1** Mark the turning point and y intercept (•).

**2** Turning point moved **2 units to right**
$$\Rightarrow y = a(\boldsymbol{x - 2})^2 \pm v$$

**3** Turning point moved **3 units down**
$$\Rightarrow y = a(x - 2)^2 - \boldsymbol{3}$$

**4** Substitute the coordinates of the
y intercept, (0, −5):
$$-5 = a(0 - 2)^2 - 3$$
$$-2 = 4a$$
$$a = -\frac{1}{2}$$

So the equation must be $y = -\frac{1}{2}(x - 2)^2 - 3$.

ISBN: 9780170477680   PHOTOCOPYING OF THIS PAGE IS RESTRICTED UNDER LAW.

Write equations for the following parabolas.

**9**

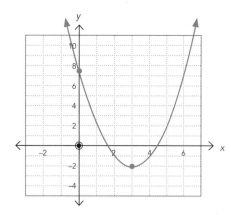

1 Mark the turning point and y intercept (•).

2 Turning point moved _____ unit(s) to _____

$\Rightarrow y = a(x\underline{\hspace{1cm}})^2 \pm v$

3 Turning point moved _____ unit(s) up/down

$\Rightarrow y = a(x\underline{\hspace{1cm}})^2 \underline{\hspace{1cm}}$

4 Substitute the coordinates of the y intercept,

(0, ____)

So the equation must be y = _____

**10**

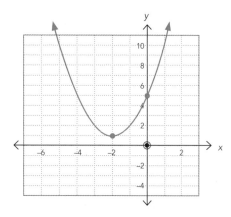

1 Mark the turning point and y intercept (•).

2 Turning point moved _____ unit(s) to _____

$\Rightarrow y = a(x\underline{\hspace{1cm}})^2 \pm v$

3 Turning point moved _____ unit(s) up/down

$\Rightarrow y = a(x\underline{\hspace{1cm}})^2 \underline{\hspace{1cm}}$

4 Substitute the coordinates of the y intercept,

(0, ____)

So the equation must be y = _____

**11**

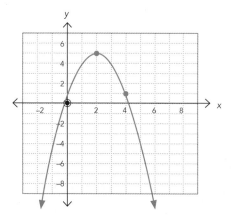

_____

_____

_____

_____

_____

_____

_____

**12**

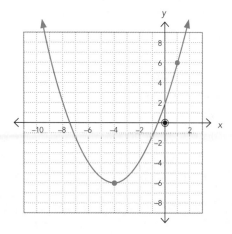

_____

_____

_____

_____

_____

_____

_____

PHOTOCOPYING OF THIS PAGE IS RESTRICTED UNDER LAW. ISBN: 9780170477680

**13**

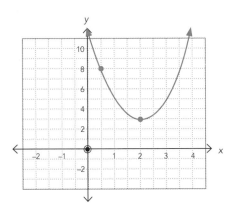

_____
_____
_____
_____
_____
_____
_____

**14**

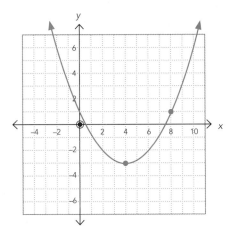

_____
_____
_____
_____
_____
_____
_____

**15**

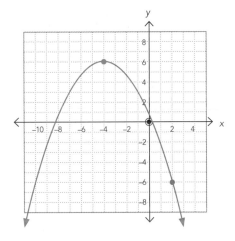

_____
_____
_____
_____
_____
_____
_____

**16**

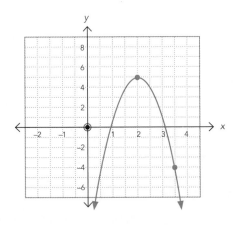

ISBN: 9780170477680    PHOTOCOPYING OF THIS PAGE IS RESTRICTED UNDER LAW.

## Quadratic inequations

**Examples:**

**1** For what range of values of $x$ is $x^2 - 8x + 15 > 0$?

Solve the equation $x^2 - 8x + 15 = 0$

$$(x - 3)(x - 5) = 0 \Rightarrow x = 3 \text{ or } x = 5$$

Sketch the graph of $(x - 3)(x - 5) = 0 \therefore x^2 - 8x + 15 > 0$, where $x < 3$ and $x > 5$.

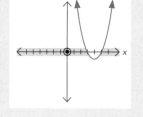

**2** Solve $8 < \dfrac{x^2 - 1}{3} < 16$

$$24 < x^2 - 1 < 48$$

$$25 < x^2 < 49$$

$\therefore$ is a short way of saying '**therefore**'.

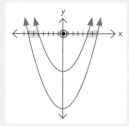

Sketch the graphs of $x^2 - 25 = 0$ and $x^2 - 49 = 0 \therefore 8 < \dfrac{x^2 - 1}{3} < 16$, where $5 < x < 7$ and $-7 < x < -5$.

53

Answer the following questions, and show your reasoning.

**1** For what range of values of $x$ is $x^2 + 3x - 10 \leq 0$?

**2** For what values will p be negative for the following relationship? $p = x^2 - 5x - 14$

**3** Solve $17 < \dfrac{x^2 + 4}{5} < 25$.

**4** Solve $5 < \dfrac{x^2 + 6}{2} < 53$.

PHOTOCOPYING OF THIS PAGE IS RESTRICTED UNDER LAW. ISBN: 9780170477680

# Mixing it up

Answer the following.

1 A graph with the equation $y = 3(x + 2)(x - 4)$ is wider/skinnier than a graph with the equation $y = (x + 2)(x - 4)$.

_____

2 A graph with the equation $y = \dfrac{(x - 5)^2}{2} + 3$ is wider/skinnier than a graph with the equation $y = (x + 1)^2 + 7$.

_____

3 Write the coordinates of the point(s) where $y = x(x - 5)$

    **a** cuts the $x$-axis _____

    **b** cuts the $y$-axis _____

4 For the graph of $y = (x + 3)(2x - 7)$, write the

    **a** $x$ intercept(s) _____

    **b** $y$ intercept(s) _____

5 Write down the turning point of the graph of $y = -(x - 6)^2 - 1$. _____

This point is a maximum/minimum.

6 Find the turning point of the graph of $y = x^2 + 6x + 8$.

_____

This point is a maximum/minimum.

7 Write a quadratic equation that has a minimum at $(1, -5)$ and a $y$ intercept of $-4$.

_____

_____

8 Write a quadratic equation that has a $y$ intercept of 2 and a maximum at $(-1, 3)$.

_____

_____

9 Find the values of p and q that will make $(2x + p)^2 = 4x^2 + qx + 25$ true, given that both p and q are positive numbers.

_____

_____

10 Find a value for c so that $y = x^2 + 8x + c$ has exactly one solution.

_____

11 Find a value for b so that $y = x^2 + bx + 81$ touches the $x$-axis at one point only.

_____

**12** Fill in the blanks with positive integers so these quadratic expressions can be factorised.

    **a**    $x^2 +$ _____ $x + 49$          **b**    $x^2 +$ _____ $x - 15$

    **c**    $x^2 -$ _____ $x - 11$          **d**    $2x^2 +$ _____ $x - 3$

**13**    **a**    Write the equation of this graph.

        _____

    **b**    Change one part of the equation so that the graph touches the x-axis at one point only.

        _____

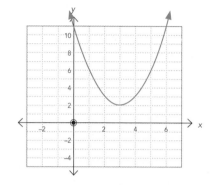

**14**    **a**    Write the equation of this graph.

        _____

    **b**    Change one part of the equation so that the x intercepts become −1 and 5.

        _____

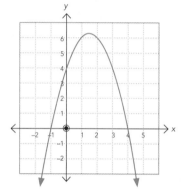

**15**    **a**    Write the y intercept for the graph of $y = (x + 2)(x + 1)$.      _____

    **b**    Change one part of the equation so that its x intercepts stay the same, but the y intercept becomes +6.      _____

**16**    **a**    Write the y intercept for the graph of $y = (x + 2)(x - 6)$.      _____

    **b**    Change one part of the equation so that its x intercepts stay the same, but the y intercept becomes −6.      _____

**17**    If $y = (x + 5)(3 - x)$, for what values of x will y be negative?

_____

_____

**18**    If $y = x^2 - 5x - 14$, for what values of x will y be negative?

_____

_____

PHOTOCOPYING OF THIS PAGE IS RESTRICTED UNDER LAW.    ISBN: 9780170477680

 Challenge 2

Answer the following.

**1**    Solve the inequality $8x^2 + 5x - 21 \leq 9 + 3x^2$

_____

_____

_____

**2**    Solve $\dfrac{3}{x^2 + 2x + 10} = \dfrac{5}{x^2 + 14}$

_____

_____

_____

**3**    Solve $\dfrac{x(x + 3)}{x^2 + 3x + 5} = \dfrac{2}{3}$

_____

_____

_____

**4**    The equation $7 = 2x^2 - 13x$ has two solutions, m and n, with m being greater than n. What is the value of m – n?

_____

_____

**5**    Show that the area of this shape can be written as $2x^2 + 21x + 34$.

_____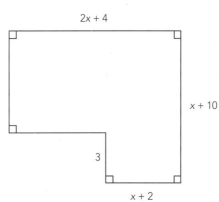

_____

_____

_____

**6**   For what values of x is $(x - 3)(x + 3) > (x - 3)(x + 4)$?

_____

_____

_____

**7**   $A = 3(x^2 - 5x + 2) + 3x$, $B = (2x + 2)(x - 7) + x^2 - 1$. Write an expression for A in terms of B.

_____

_____

_____

_____

_____

_____

_____

**8**   The graph of $ax^2 + bx + 3$ passes through the points (2, −3) and (10, 13). Find the values of a and b, and write the equation for the parabola.

_____

_____

_____

_____

_____

_____

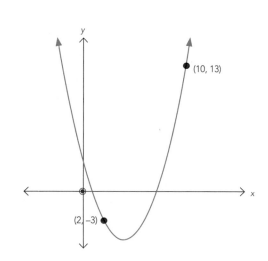

**9**   Write $\dfrac{9x^2 - 4}{3(3x + 2)} \times \dfrac{2x}{x^2 - 3x}$ in the form $\dfrac{ax + b}{cx + d}$, where a, b, c, d are integers.

_____

_____

_____

_____

_____

_____

PHOTOCOPYING OF THIS PAGE IS RESTRICTED UNDER LAW.   ISBN: 9780170477680

# Exponential equations

## Basic exponential equations

- The terms **power**, **exponent** and **index** all mean the **same**.
- In exponential equations, the unknown is the **exponent** (or power or index).

Remember:

Base — **2**$^x$ — Power, exponent or index

### Solving exponential equations

The easiest way for you to solve these is by listing the powers of the base.

**Examples:**

**1** $2^x = 32$

List the powers of 2:

| |
|---|
| $2^0 = 1$ |
| $2^1 = 2$ |
| $2^2 = 4$ |
| $2^3 = 8$ |
| $2^4 = 16$ |
| $2^5 = 32$ |

If the base numbers are the same, the powers must be equal.

$\therefore 2^x = 2^5$

So $x = 5$

**2** $3^x = 81$

List the powers of 3:

| |
|---|
| $3^0 = 1$ |
| $3^1 = 3$ |
| $3^2 = 9$ |
| $3^3 = 27$ |
| $3^4 = 81$ |

$\therefore 3^x = 3^4$

So $x = 4$

 54

Solve the following exponential equations.

**1** $5^x = 625$

_____

_____

_____

**2** $3^x = 243$

_____

_____

_____

**3** $7^x = 343$

_____

_____

_____

**4** $4^x = 16\ 384$

_____

_____

_____

**5** $2^x = 2048$

_____

_____

_____

**6** $9^x = 1$

_____

_____

_____

**7** $10^x = 100\ 000$

_____

_____

_____

**8** $0.1^x = 0.001$

_____

_____

_____

## Plotting exponential graphs

The basic exponential graph is written **y = aˣ**.

**Example:** $y = 2^x$

| Term number (x) | 0 | 1 | 2 | 3 | 4 | 5 | 6 |
|---|---|---|---|---|---|---|---|
| $2^x$ | $2^0$ | $2^1$ | $2^2$ | $2^3$ | $2^4$ | $2^5$ | $2^6$ |
| y | 1 | 2 | 4 | 8 | 16 | 32 | 64 |

− 1    + 2    + 4    + 8    + 16    + 32

If we extend the pattern to the left, we find that $2^0 = 1$.

Once again, notice that, unlike linear patterns, the increase from term to term is **not constant**.

Regardless of the value of the base, it is always true that $a^0 = 1$.

Notice that if we use the same scale on both axes, the graph appears very tall and narrow.

So, generally when we draw graphs of exponential patterns, we use different scales on the x and y axes.

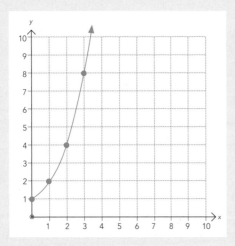

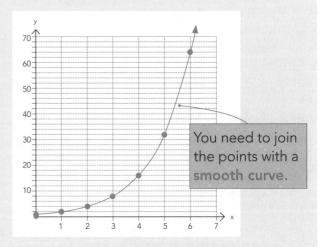

You need to join the points with a smooth curve.

### Increasing the base

Use the $x^{\blacksquare}$ button on your calculator to help you complete the table.

| Term number (x) | 0 | 1 | 2 | 3 | 4 | 5 | 6 |
|---|---|---|---|---|---|---|---|
| $2^x$ | 1 | 2 | 4 | 8 | 16 | 32 | 64 |
| $3^x$ | 1 | | | | 81 | | |
| $5^x$ | | | | 125 | | | |
| $10^x$ | | | | | | 100 000 | |

A few of these values will fit on the graph on the next page. Check that these agree with the points on the curves.

PHOTOCOPYING OF THIS PAGE IS RESTRICTED UNDER LAW.    ISBN: 9780170477680

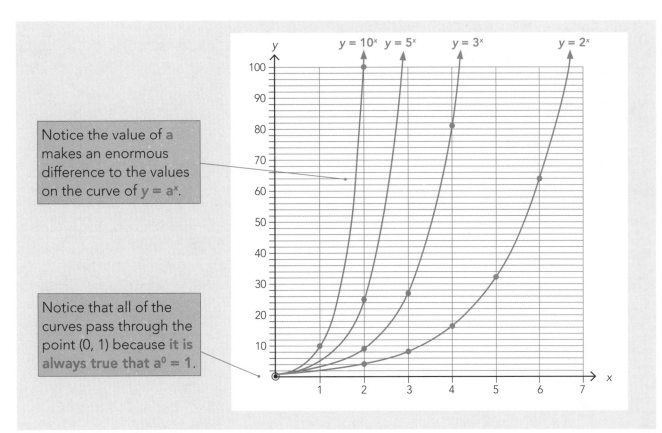

Notice the value of **a** makes an enormous difference to the values on the curve of $y = a^x$.

Notice that all of the curves pass through the point (0, 1) because **it is always true that $a^0 = 1$**.

55

Complete the table, plot the points and connect them to make a smooth curve.

**1**    $y = 4^x$

| x | y |
|---|---|
| 3 | |
| 2 | |
| 1 | |
| 0 | |
| −1 | |
| −2 | |
| −3 | |

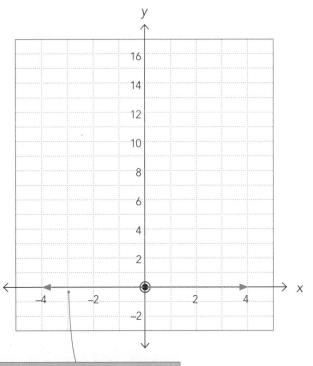

Your curve will approach this line and get really close to it, but never actually touch it. It is called the **asymptote** to the curve.

## Translated exponential graphs

- The equation takes the form $y = a^{x \pm h} \pm v$, where a = base, h = horizontal shift, v = vertical shift.

### Vertical shift

$y = 2^x + 4$      Means the graph is translated **4 units up**.
$y = 2^x - 2$      Means the graph is translated **2 units down**.

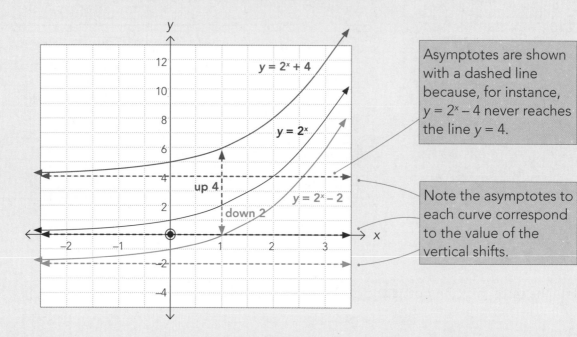

Asymptotes are shown with a dashed line because, for instance, $y = 2^x - 4$ never reaches the line $y = 4$.

Note the asymptotes to each curve correspond to the value of the vertical shifts.

### Horizontal shift

$y = 2^{x+1}$      Means the graph is translated **1 unit to the left**.
$y = 2^{x-3}$      Means the graph is translated **3 units to the right**.

Notice that these move in the **opposite direction** from what you might expect.

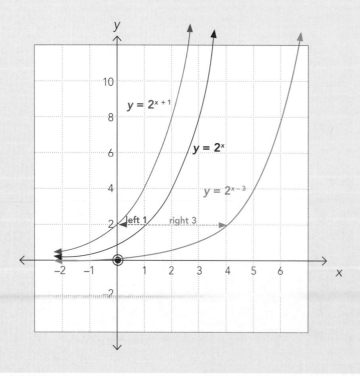

Notice that they all have asymptotes on the x-axis.

PHOTOCOPYING OF THIS PAGE IS RESTRICTED UNDER LAW. ISBN: 9780170477680

Match these equations to the correct graphs.

| $y = 2^x + 2$ | $y = 2^{x-1}$ | $y = 2^x - 1$ |
|---|---|---|
| $y = 2^x + 3$ | $y = 2^{x-2}$ | $y = 2^{x+2}$ |

**1**

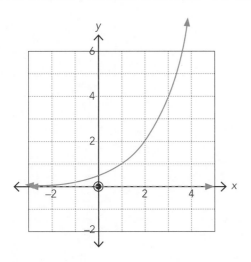

_____

**2**

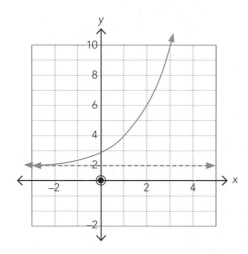

_____

**3**

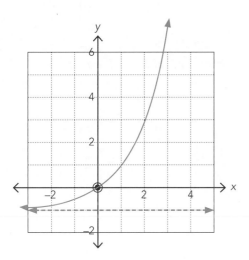

_____

**4**

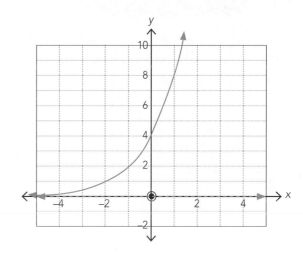

_____

**5**

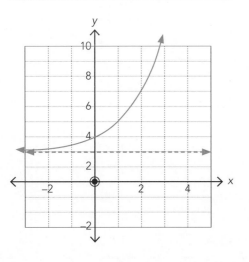

_____

**6**

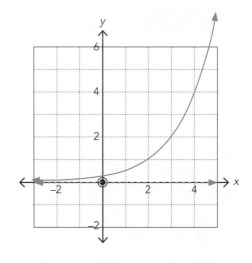

_____

Complete the table, plot the points and connect them to make a smooth curve.

**7**  $y = 3^x - 2$

| x | y |
|---|---|
| 0 |   |
| 1 |   |
| 2 |   |

The y intercept is at (_____ ,_____)

The equation of the asymptote is

$y =$ _____

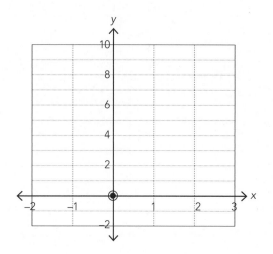

**8**  $y = 4^{x+1}$

| x | y |
|---|---|
| −1 |   |
| 0 |   |
| 1 |   |

The y intercept is at (_____ ,_____)

The equation of the asymptote is

_____

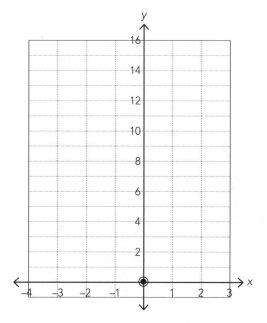

**9**  $y = 2^{0.5x}$

| x | y |
|---|---|
| −2 |   |
| 0 |   |
| 2 |   |
| 4 |   |
| 6 |   |

The y intercept is at (_____ ,_____)

The equation of the asymptote is

_____

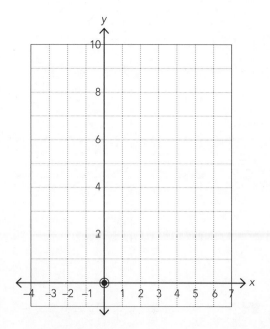

PHOTOCOPYING OF THIS PAGE IS RESTRICTED UNDER LAW.    ISBN: 9780170477680

# Challenge 3

Solve the following equations and inequations.

**1**   $2^{x^2} \leq 2^{6x}$

_____

_____

_____

_____

_____

**2**   $2^{x^2} = 8 \div 2^{2x}$

_____

_____

_____

_____

_____

**3**   $2^{x^2} > 2^{3x} \times 4^5$

_____

_____

_____

_____

_____

**4**   $3^{x^2} \times 3^{12x} \div 27 > 0$

_____

_____

_____

_____

_____

Answer the following questions.

**5**   If $n$ is a whole number, for what values of $n$ is $7 \times 2^{n+3} < 400$?

_____

_____

_____

_____

**6**   If $n$ is a whole number, for what values of $n$ is $11 \times 3^{2n-5} > 500$?

_____

_____

_____

_____

**7**   For what values of $n$ will $5^{3n} = 25 \times 5^{-2n^2}$?

_____

_____

_____

_____

_____

**8**   For what values of $p$ will $256 \times 4^{p^2} = 16^{2p}$?

_____

_____

_____

_____

_____

# Rearrangement of expressions

## 1 Where the subject appears once

- First, put **all** terms which contain the required subject on the left.
- Get rid of fractions by multiplying by the lowest common multiple of the denominators.
- Get rid of square roots by isolating the square root on one side and then squaring both sides.
- Use normal equation-solving rules to isolate the subject.

**Examples:**

**1** Make $r$ the subject of $V = \pi r^2 h$.

$$\pi r^2 h = V \qquad (\div \pi h)$$
$$r^2 = \frac{V}{\pi h} \qquad (\sqrt{\ } \text{ both sides})$$
$$r = \pm\sqrt{\frac{V}{\pi h}}$$

**2** Express $x$ in terms of $y$ if $(x^4)^5 = (y^2)^5$.

$$(x^4)^5 = (y^2)^5 \qquad \text{(multiply indices)}$$
$$x^{20} = y^{10}$$
$$x^{20} = (y^{\frac{1}{2}})^{20}$$
$$x = \pm y^{\frac{1}{2}} \text{ or } \pm\sqrt{y}$$

**3** Rewrite the formula $A = \frac{(a+b)}{2}h$ with $b$ as the subject.

$$\frac{(a+b)}{2}h = A \qquad (\times 2)$$
$$(a+b)h = 2A \qquad \text{(expand brackets)}$$
$$ah + bh = 2A \qquad (-ah)$$
$$bh = 2A - ah \qquad (\div h)$$
$$b = \frac{2A - ah}{h}$$

**4** Make $y$ the subject of the formula.

$$x = \frac{\sqrt{y^3 + 2}}{5} \qquad \text{(isolate term with } \sqrt{\ })$$
$$\sqrt{y^3 + 2} = 5x \qquad \text{(square both sides)}$$
$$y^3 + 2 = 25x^2 \qquad (-2)$$
$$y^3 = 25x^2 - 2 \qquad (\sqrt[3]{\ } \text{ both sides})$$
$$y = \sqrt[3]{25x^2 - 2}$$

Rearrange the following equations.

**1** Make $c$ the subject of $a = \sqrt{2b - c}$.

**2** Make $R$ the subject of $I = \frac{PRT}{100}$.

**3** Make $y$ the subject of $x = \sqrt{\frac{3y}{2}}$.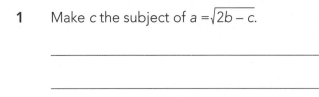

**4** Make $b$ the subject of $V = \frac{1}{3}b^2 h$.

PHOTOCOPYING OF THIS PAGE IS RESTRICTED UNDER LAW.   ISBN: 9780170477680

**5**  Rearrange $y = mx + c$ so that $m$ is the subject.

_____

_____

_____

_____

**6**  Make $a$ the subject of $v^2 = u^2 + 2as$.

_____

_____

_____

_____

**7**  The formula for converting degrees Fahrenheit into degrees Celsius is $F = 32 + \dfrac{9C}{5}$. Make $C$ the subject.

_____

_____

_____

_____

_____

**8**  Make $r$ the subject of the formula for the volume of a sphere: $V = \dfrac{4}{3}\pi r^3$.

_____

_____

_____

_____

_____

**9**  Make $\theta$ the subject of the formula for the area of a sector: $A = \pi r^2 \times \dfrac{\theta}{360}$.

_____

_____

_____

_____

_____

**10**  Make $u$ the subject of the equation $s = ut + \dfrac{1}{2}at^2$.

_____

_____

_____

_____

_____

**11**  Express $a$ in terms of $b$ if $(a^4)^3 \times a^2 = b(b^3)^2$.

_____

_____

_____

_____

**12**  Express $a$ in terms of $b$ if $\dfrac{(a^3)^2}{a} = (b^5)^2$.

_____

_____

_____

_____

## 2 Where the subject appears twice

- If the subject appears in several terms, collect all of these to one side and then factorise.

**Examples:**

**1** Make $c$ the subject of $a = \dfrac{16(b - 3c)}{c}$

$$a = \dfrac{16(b - 3c)}{c} \quad (\times c)$$

$$ac = 16(b - 3c) \quad (\text{expand})$$

$$ac = 16b - 48c \quad (+ 48c)$$

$$ac + 48c = 16b \quad (\text{factorise})$$

$$c(a + 48) = 16b \quad (\div (a + 48))$$

$$c = \dfrac{16b}{a + 48}$$

Get both terms with $c$ to the left.

**2** Make $e$ the subject of $\dfrac{2}{c} = \dfrac{3}{e} + \dfrac{7}{d}$

$$\dfrac{2}{c} = \dfrac{3}{e} + \dfrac{7}{d} \quad (\times \dfrac{ced}{1})$$

$$\dfrac{2}{c} \times \dfrac{ced}{1} = \dfrac{3}{e} \times \dfrac{ced}{1} + \dfrac{7}{d} \times \dfrac{ced}{1} \quad (\text{simplify})$$

$$2ed = 3cd + 7ce \quad (- 7ce)$$

$$2ed - 7ce = 3cd \quad (\text{factorise})$$

$$e(2d - 7c) = 3cd \quad (\div (2d - 7c))$$

$$e = \dfrac{3cd}{2d - 7c}$$

58

Rearrange the following equations.

**1** Make $b$ the subject of this equation:

$8b = 12 - ab$

_____

_____

_____

_____

_____

**2** Make $y$ the subject of this equation:

$2(x - y) = 5y + 9$

_____

_____

_____

_____

_____

**3** Make $a$ the subject of this equation:

$b + 3 = \dfrac{7}{a}$

_____

_____

_____

_____

_____

**4** Make $b$ the subject of this equation:

$3(4b - a) = 6a(2 - b)$

_____

_____

_____

_____

_____

PHOTOCOPYING OF THIS PAGE IS RESTRICTED UNDER LAW.   ISBN: 9780170477680

**5** Make $y$ the subject of this equation:

$$\frac{2y + 3}{x} = y + 5$$

_____

_____

_____

_____

_____

_____

**6** Make $b$ the subject of this equation:

$$a = \frac{2b + 5}{b - 1}$$

_____

_____

_____

_____

_____

_____

**7** Show that the equation $\frac{1}{c} + \frac{1}{b} = \frac{1}{a}$ can be

rearranged to give $b = -\frac{ac}{c - a}$

_____

_____

_____

_____

_____

_____

**8** Show that the equation $\frac{4}{a} = 3 - \frac{2}{b}$ can be

rearranged to give $b = -\frac{2a}{4 - 3a}$

_____

_____

_____

_____

_____

_____

**9** Give the equation for $a$ in terms of $b$, $c$ and $d$.

$$\frac{a}{a + b} = \frac{c}{d}$$

_____

_____

_____

_____

_____

**10** Give the equation for $b$ in terms of $a$ and $c$.

$$a = \frac{c - b}{3b - 7}$$

_____

_____

_____

_____

_____

# Geometry and space

## Fundamentals

### Labelling angles and sides in triangles

- Vertices are usually given upper-case letters and sides are given lower-case letters.

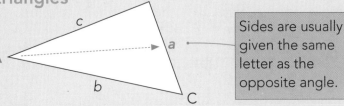

Sides are usually given the same letter as the opposite angle.

- Angles can be named in two ways.

**1 Using the three letters outside each vertex.** This angle is $\angle DEF$ or $\angle FED$.

**2 Using the one letter inside the angle.** This angle is $\angle y$.

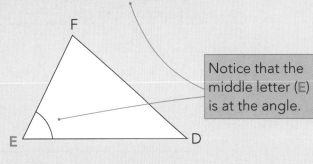

Notice that the middle letter (E) is at the angle.

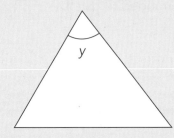

### Basic angles and reasons

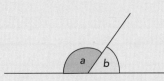

$a + b = 180°$
Angles on a line add to 180°
($\angle$s on a line $\Rightarrow$ 180°)

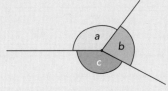

$a + b + c = 360°$
Angles at a point add to 360°
($\angle$s at a point $\Rightarrow$ 360°)

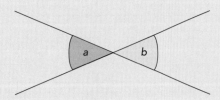

$a = b$
Vertically opposite angles are equal
(vert opp $\angle$s =)

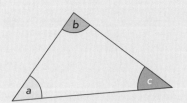

$a + b + c = 180°$
Angles in a triangle add to 180°
($\angle$s in $\Delta \Rightarrow$ 180°)

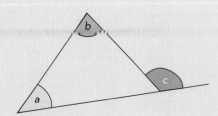

$a + b = c$
The exterior angle of a triangle = the sum of the interior opposite angles.
(ext $\angle$ of $\Delta$ = sum of int opp $\angle$s)

PHOTOCOPYING OF THIS PAGE IS RESTRICTED UNDER LAW.    ISBN: 9780170477680

## Parallel lines

| Relationship | Reason |
|---|---|
| $a = b$ | Alternate angles on parallel lines are equal. (These form a **'Z'**.)<br><br>(alt $\angle$s =, // lines) |
| $a = b$ | Corresponding angles on parallel lines are equal. (These form an **'F'**.)<br><br>(corr $\angle$s =, // lines) |
| $a + b = 180°$ | Co-interior angles on parallel lines add to 180°. (These form a **'C'**.)<br><br>(co-int $\angle$s add to 180°, // lines) |

## Polygons

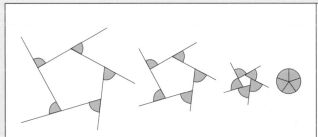

Exterior angles of a polygon add to 360°.

(ext $\angle$s of polygon = 360°)

**Example:**
Hexagon

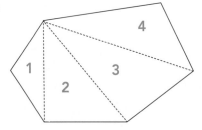

Interior angles add to 4 x 180° = 720°

Interior angles of a polygon add to
$(n - 2) \times 180°$.
Each interior angle of a **regular** polygon =
$$\frac{(n - 2) \times 180°}{n} = 180° - \frac{360°}{n}$$

ISBN: 9780170477680    PHOTOCOPYING OF THIS PAGE IS RESTRICTED UNDER LAW.

# Right-angled triangles

## Theorem of Pythagoras

- The Theorem of Pythagoras applies to **right-angled** triangles only.
- The longest side is the **hypotenuse**.
- The hypotenuse is always **opposite** the right angle.
- The theorem is used for finding **lengths** of sides.

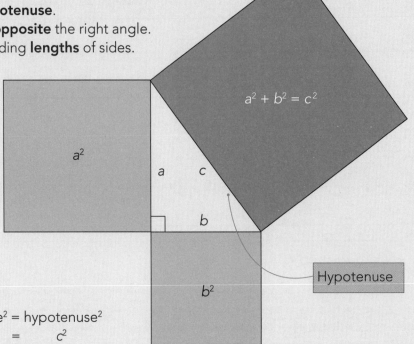

$$a^2 + b^2 = c^2$$

Hypotenuse

short side$^2$ + short side$^2$ = hypotenuse$^2$
$$a^2 \quad + \quad b^2 \quad = \quad c^2$$

**Examples:**

**1 Finding the length of the hypotenuse**
Calculate the length of c.

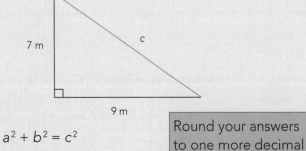

$$a^2 + b^2 = c^2$$
$$7^2 + 9^2 = c^2$$
$$\sqrt{7^2 + 9^2} = c$$
$$c = \sqrt{130}$$
$$c = 11.4 \text{ m (1 dp)}$$

Round your answers to one more decimal place than the lengths given.

11.4 m is a reasonable answer because it's longer than the other two sides.

**2 Finding the length of short sides**
Calculate the length of y.

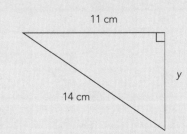

$$a^2 + b^2 = c^2$$
$$y^2 + 11^2 = 14^2$$
$$y^2 = 14^2 - 11^2$$
$$y = \sqrt{14^2 - 11^2}$$
$$y = \sqrt{75}$$
$$y = 8.7 \text{ cm (1 dp)}$$

8.7 cm is a reasonable answer because it's shorter than the hypotenuse.

PHOTOCOPYING OF THIS PAGE IS RESTRICTED UNDER LAW.     ISBN: 9780170477680

Calculate the unknown length of each triangle. Round your answers to one more decimal place than the lengths given.

**1**

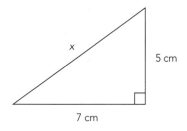

x
5 cm
7 cm

_____

_____

_____

**2**

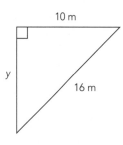

10 m
y
16 m

_____

_____

_____

**3**

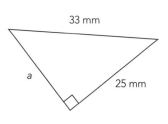

33 mm
a
25 mm

_____

_____

_____

**4**

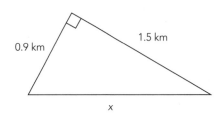

0.9 km
1.5 km
x

_____

_____

_____

**5**

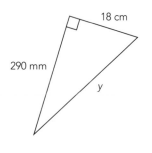

18 cm
290 mm
y

_____

_____

_____

**6**

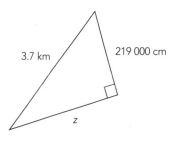

3.7 km
219 000 cm
z

_____

_____

_____

**7** This equilateral triangle has sides of 1.9 m.

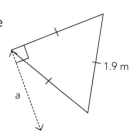

1.9 m
a

_____

_____

_____

**8** This isosceles triangle has a vertical height of 35 cm.

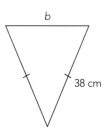

b
38 cm

_____

_____

_____

ISBN: 9780170477680     PHOTOCOPYING OF THIS PAGE IS RESTRICTED UNDER LAW.

## Multistep calculations

- When you need to do several calculations to reach an answer, round intermediate answers to at least one more decimal place than what is needed for the final answer.

- It is better still if you can use your 'Answer' button  on your calculator so you **don't** round at all until you reach the final answer.

**Example:** Calculate the length of AD.

Step 1: BD is common to both triangles, so calculate its length.

$$BD^2 = CD^2 + BC^2$$
$$BD = \sqrt{91^2 + 48^2}$$
$$BD = 102.88 \text{ mm}$$

Step 2: Use the length of BD to calculate the length of AD.

$$AD^2 = BD^2 - AB^2$$
$$AD = \sqrt{102.88^2 - 66^2}$$
$$AD = 78.9 \text{ mm (1 dp)}$$

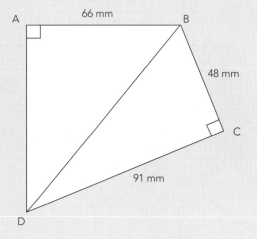

Answer the following questions.

**9** Calculate the length of BC.

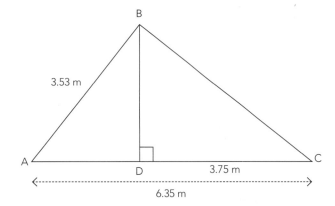

_____

_____

_____

_____

_____

_____

_____

_____

_____

**10** Calculate the length of AB.

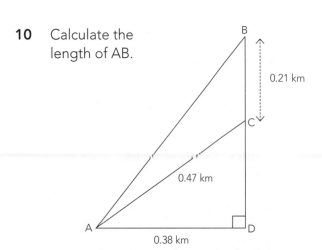

_____

_____

_____

_____

_____

_____

_____

PHOTOCOPYING OF THIS PAGE IS RESTRICTED UNDER LAW. ISBN: 9780170477680

## Pythagoras and algebra

### Examples:

**1** Calculate the missing dimensions of all three sides of this right-angled triangle.

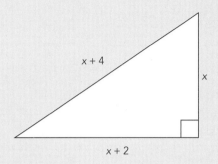

Use the Theorem of Pythagoras to write an algebraic equation, then solve it.

$$x^2 + (x + 2)^2 = (x + 4)^2$$
$$x^2 + (x + 2)(x + 2) = (x + 4)(x + 4)$$
$$x^2 + (x^2 + 4x + 4) = x^2 + 8x + 16$$
$$x^2 - 4x - 12 = 0$$
$$(x + 2)(x - 6) = 0$$
$$x = -2 \text{ or } 6$$

> It is important that you **check** whether both answers are possible. Usually a quadratic equation has two solutions and often one doesn't work in practice.

The side of a triangle cannot be –2, so x must be 6.
Therefore the sides are 6, 8 and 10.

**2** Calculate the length of x.

> State your reasoning.

BD is common to both triangles.

So: 
$$28^2 - x^2 = 26^2 - (40 - x)^2$$
$$28^2 - x^2 = 26^2 - (1600 - 80x + x^2)$$
$$28^2 = 26^2 - 1600 + 80x$$
$$x = \frac{28^2 - 26^2 + 1600}{80}$$
$$x = 21.35 \text{ cm}$$

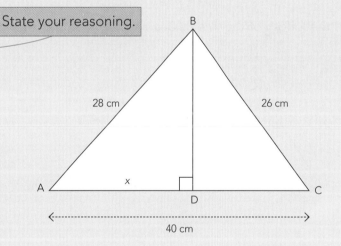

Answer the following questions.

**11** If the lengths of the two shorter sides of a right-angled triangle are x and (x + 4), write an expression for the length of the hypotenuse.

_____

_____

_____

_____

**12** If the lengths of the two shorter sides of a right-angled triangle are ($x - 3$) and ($x + 3$), write an expression for the length of the hypotenuse.

_____

_____

_____

_____

**13** If the length of the hypotenuse of a right-angled triangle is ($4x + 1$) and the shortest side is ($x + 2$), write an expression for the length of the third side.

_____

_____

_____

_____

**14** The vertical height of a cone is three times the length of its radius ($r$). Write an expression for the slant height of the cone ($s$).

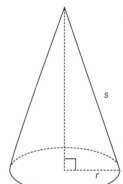

_____

_____

_____

Find the length of the three sides of these triangles.

**15**

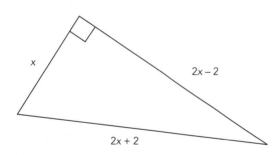

_____

_____

_____

_____

_____

**16**

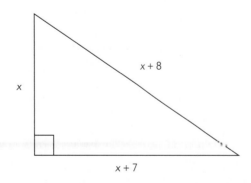

_____

_____

_____

_____

_____

 PHOTOCOPYING OF THIS PAGE IS RESTRICTED UNDER LAW.   ISBN: 9780170477680

**17**

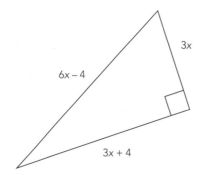

**18**

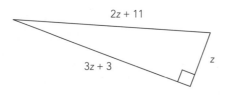

**19** Calculate the value of **y**.

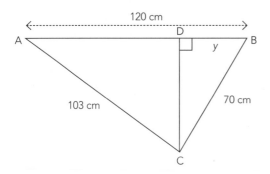

**20** This figure represents a cuboid.
EH = 4**x**, GH = 3**x** and CG = 12**x**.
Calculate its dimensions.

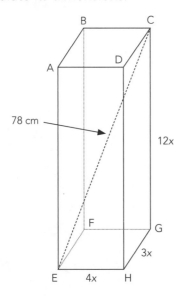

## Trigonometry

- Trigonometry can also be used for calculations involving **right-angled triangles**.
- However, unlike calculations using the Theorem of Pythagoras, an **angle must** be involved.
- Before you do any calculations, make sure your calculator is in **degree** mode.

Before you begin a calculation involving trigonometry, **label** the sides of the triangle:

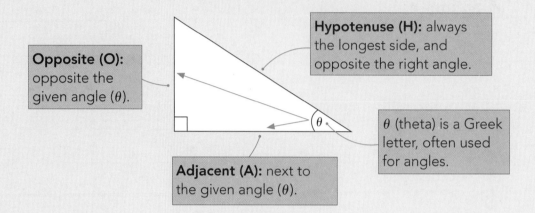

**Opposite (O):** opposite the given angle ($\theta$).

**Hypotenuse (H):** always the longest side, and opposite the right angle.

$\theta$ (theta) is a Greek letter, often used for angles.

**Adjacent (A):** next to the given angle ($\theta$).

## SOH CAH TOA

You will need to know about three trigonometrical functions: **sin** $\theta$ (sine)

**cos** $\theta$ (cosine)

**tan** $\theta$ (tangent).

These are the rules in trigonometry:

$$\sin \theta = \frac{O}{H} \qquad \cos \theta = \frac{A}{H} \qquad \tan \theta = \frac{O}{A}$$

These are usually remembered as the 'word' **SOH CAH TOA**.

Organising **SOH CAH TOA** into triangles can help you to work out how to use it:

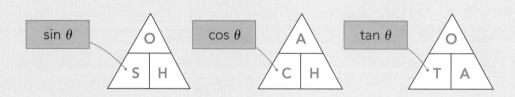

### Process

Step 1: **Label** the sides that are involved in the problem with A, O and H.

Step 2: Decide **which relationship** involves these sides.

Step 3: **Substitute** the values from the triangle.

Step 4: Include **units** for angles (°) and sides (cm, etc.).

Step 5: **Think about your answer — does it seem sensible?**

PHOTOCOPYING OF THIS PAGE IS RESTRICTED UNDER LAW.  ISBN: 9780170477680

### Finding sides

- We can find an unknown side of a right-angled triangle using trigonometry if we have an angle (other than the right angle) and another side.

#### Using sine
**Examples:**

**1** Finding a **short** side (O) using **sine**.

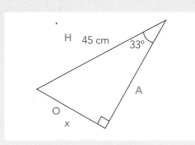

$$\sin \theta = \frac{O}{H}$$

$$\sin 33° = \frac{x}{45}$$

$$x = 45 \times \sin 33°$$

$$= 24.5 \text{ cm (1 dp)}$$

24.5 cm seems a reasonable answer because it's shorter than the hypotenuse.

**2** Finding the **hypotenuse** (H) using **sine**.

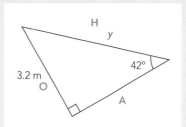

$$\sin \theta = \frac{O}{H}$$

$$\sin 42° = \frac{3.2}{y}$$

$$y \times \sin 42° = 3.2$$

$$y = \frac{3.2}{\sin 42°}$$

$$y = 4.78 \text{ m (2 dp)}$$

4.78 m seems a reasonable answer because it's the hypotenuse and it's longer than the opposite side.

#### Using cosine
**Examples:**

**1** Finding a **short** side (A) using **cosine**.

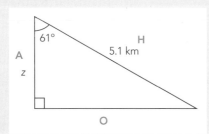

$$\cos \theta = \frac{A}{H}$$

$$\cos 61° = \frac{z}{5.1}$$

$$z = 5.1 \times \cos 61°$$

$$= 2.47 \text{ km (2 dp)}$$

2.47 km seems a reasonable answer because it's shorter than the hypotenuse.

**2** Finding the **hypotenuse** (H) using **cosine**.

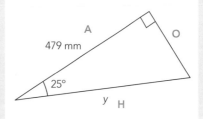

$$\cos \theta = \frac{A}{H}$$

$$\cos 25° = \frac{479}{y}$$

$$y \times \cos 25° = 479$$

$$y = \frac{479}{\cos 25°}$$

$$y = 528.5 \text{ mm (1 dp)}$$

528.5 mm seems a reasonable answer because it's the hypotenuse and it's longer than the adjacent side.

**Using tangent**
**Examples:**

**1** Finding an **opposite** side (O) using **tangent**.

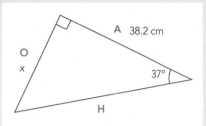

$$\tan \theta = \frac{O}{A}$$

$$\tan 37° = \frac{x}{38.2}$$

$$x = 38.2 \times \tan 37°$$

$$= 28.79 \text{ cm (2 dp)}$$

28.79 cm seems a reasonable answer because it's opposite the smallest angle and shorter than the adjacent side.

**2** Finding an **adjacent** side (A) using **tangent**.

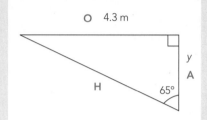

$$\tan \theta = \frac{O}{A}$$

$$\tan 65° = \frac{4.3}{y}$$

$$y \times \tan 65° = 4.3$$

$$y = \frac{4.3}{\tan 65°}$$

$$y = 2.01 \text{ m (2 dp)}$$

2.01 m seems a reasonable answer because it's opposite the smallest angle and shorter than the opposite side.

Use trigonometry to calculate the unknown length in each triangle. Write your answers to 3 significant figures.

**1**

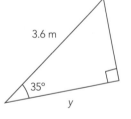

_____

_____

_____

**2**

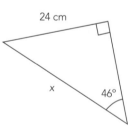

_____

_____

_____

**3**

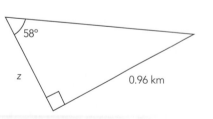

_____

_____

_____

**4**

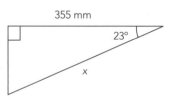

_____

_____

_____

PHOTOCOPYING OF THIS PAGE IS RESTRICTED UNDER LAW. ISBN: 9780170477680

**5**

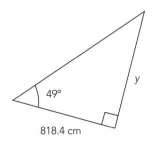

49°

818.4 cm

y

_____

_____

_____

**6**

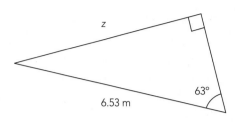

z

63°

6.53 m

_____

_____

_____

**7**

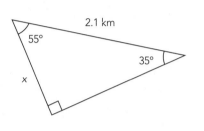

55°

2.1 km

35°

x

_____

_____

_____

**8**

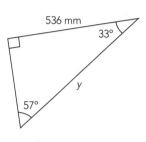

536 mm

33°

y

57°

_____

_____

_____

**9**

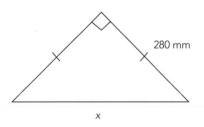

280 mm

x

_____

_____

_____

**10**

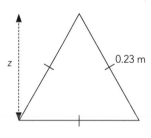

z

0.23 m

_____

_____

_____

**11**

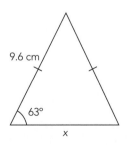

9.6 cm

63°

x

_____

_____

_____

**12**

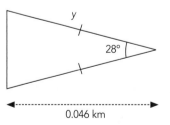

y

28°

0.046 km

_____

_____

_____

ISBN: 9780170477680     PHOTOCOPYING OF THIS PAGE IS RESTRICTED UNDER LAW.

## Finding angles

- We can find an unknown angle of a right-angled triangle using trigonometry if we know the **lengths of two sides**.
- Round angles to a maximum of 1 dp.

**Examples:**
**Using sine**

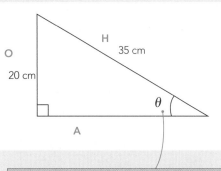

> Use the '**inverse sin**' button on your calculator.

$$\sin \theta = \frac{O}{H}$$

$$\sin \theta = \frac{20}{35}$$

'undo' sin                    'undo' sin

$$\theta = \sin^{-1}\left(\frac{20}{35}\right)$$

$$\theta = 34.8° \text{ (1 dp)}$$

> $\theta$ (theta) is a symbol that is often used for an unknown angle.

> Think about your answer — does it seem reasonable? In this case, 34.8° is reasonable because it is less than 90°, and the smallest angle is opposite the shortest side.

**Using cosine**

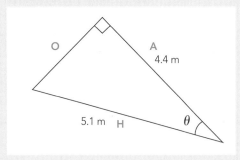

$$\cos \theta = \frac{A}{H}$$

$$\cos \theta = \frac{4.4}{5.1}$$

'undo' cos                    'undo' cos

$$\theta = \cos^{-1}\left(\frac{4.4}{5.1}\right)$$

$$\theta = 30.4° \text{ (1 dp)}$$

> Think about your answer. $\theta$ is opposite the shorter of the two perpendicular sides, so an answer less than 90° is expected.

**Using tangent**

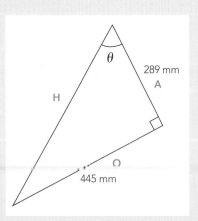

$$\tan \theta = \frac{O}{A}$$

$$\tan \theta = \frac{445}{289}$$

'undo' tan                    'undo' tan

$$\theta = \tan^{-1}\left(\frac{445}{289}\right)$$

$$\theta = 57.0° \text{ (1 dp)}$$

> Think about your answer. $\theta$ is opposite the longer of the two perpendicular sides, so an answer between 45° and 90° is expected.

PHOTOCOPYING OF THIS PAGE IS RESTRICTED UNDER LAW.     ISBN: 9780170477680

Use trigonometry to calculate the unknown angle in each triangle. Round your answers to 1 dp.

**13**

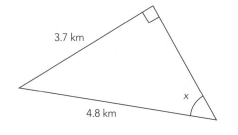

3.7 km

4.8 km

x

_____

_____

_____

**14**

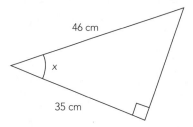

46 cm

35 cm

x

_____

_____

_____

**15**

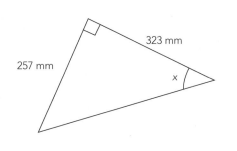

323 mm

257 mm

x

_____

_____

_____

**16**

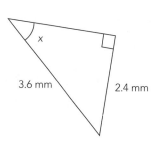

x

3.6 mm

2.4 mm

_____

_____

_____

**17**

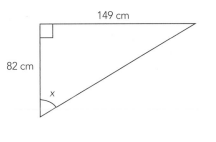

149 cm

82 cm

x

_____

_____

_____

**18**

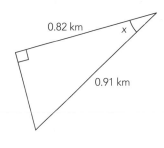

0.82 km

x

0.91 km

_____

_____

_____

**19**

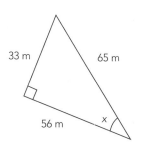

33 m

65 m

56 m

x

_____

_____

_____

**20**

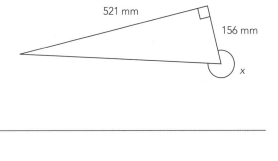

521 mm

156 mm

x

_____

_____

_____

ISBN: 9780170477680    PHOTOCOPYING OF THIS PAGE IS RESTRICTED UNDER LAW.

**Multistep calculations using trigonometry**

- As with Pythagoras problems, when you need to do several calculations to reach an answer, round intermediate answers to at least **one** more decimal place than what is given in the question.
- It is better still if you can use your 'Answer' button 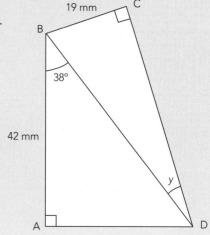 on your calculator so you don't round at all until you reach the final answer.

**Example:** Calculate angle $y$ to the nearest degree.

Step 1:   BD is common to both triangles, so calculate its length.

$$\cos 38° = \frac{42}{BD}$$

$$BD = \frac{42}{\cos 38°}$$

$$= 53.3 \text{ mm}$$

Step 2:   Use the length of BD to calculate angle $y$.

$$y = \sin^{-1}\left(\frac{19}{53.3}\right)$$

$$= 21°$$

Answer the following questions. Calculate lengths to one more decimal place than the information given and angles to the nearest degree.

**21**   AD is bisected by C. Calculate angle $x$.

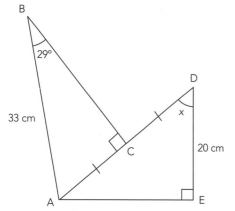

**22**   Calculate the size of side $x$.

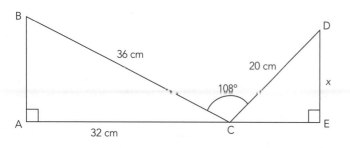

PHOTOCOPYING OF THIS PAGE IS RESTRICTED UNDER LAW.   ISBN: 9780170477680

**23** Calculate the size of angle x.

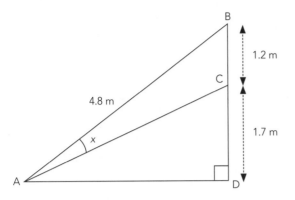

**24** Calculate the size of angle y.

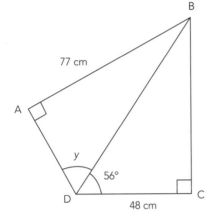

**25** Calculate the length of CD.

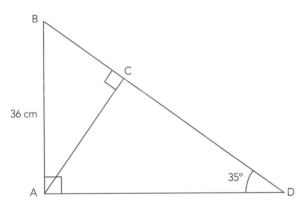

**26** Calculate the coordinates (x, y).

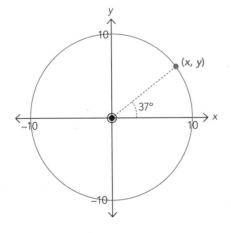

## Putting it together with Pythagoras

Which formula to use?

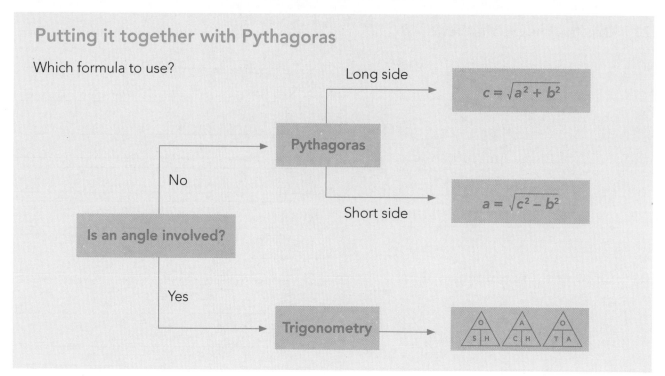

Calculate the labelled sides and angles. Give lengths to 3 sf and angles to 1 dp.

**1**

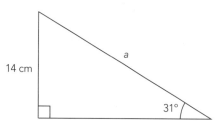

_____

_____

_____

**2**

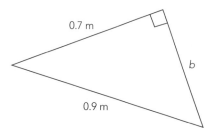

_____

_____

_____

**3**

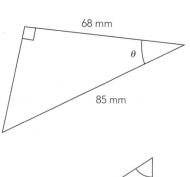

_____

_____

_____

**4**

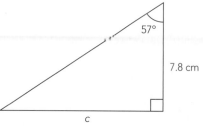

_____

_____

_____

PHOTOCOPYING OF THIS PAGE IS RESTRICTED UNDER LAW.     ISBN: 9780170477680

Answer the following questions. You will need to do more than one calculation for most answers. Remember to round intermediate answers to one more decimal place than given in the question.

**5** BE = 9 cm and DC = 32 cm. Calculate the size of angle x.

_____

_____

_____

_____

_____

_____

_____

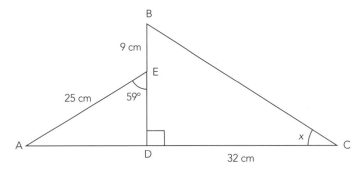

**6** BD bisects the line AC. Calculate the sizes of angles x and y.

_____

_____

_____

_____

_____

_____

_____

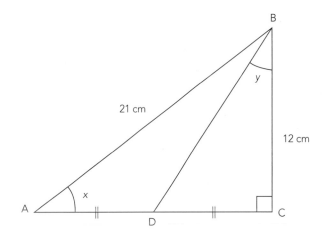

**7** Calculate the length of BC.

_____

_____

_____

_____

_____

_____

_____

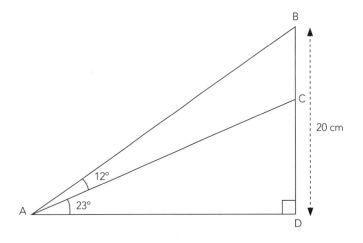

**8** Calculate the length *x*.

_____

_____

_____

_____

_____

_____

_____

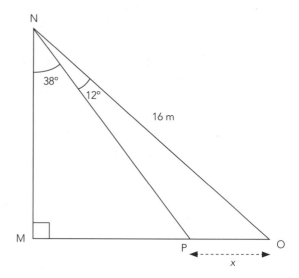

**9** Calculate the length of AD. Hint: You will need to add lines and two new points (E and F).

_____

_____

_____

_____

_____

_____

_____

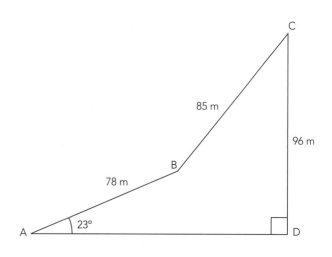

**10** Calculate length *x*. Hint: You will need to use algebra for this question.

_____

_____

_____

_____

_____

_____

_____

_____

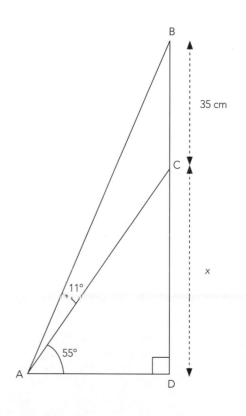

PHOTOCOPYING OF THIS PAGE IS RESTRICTED UNDER LAW.     ISBN: 9780170477680

# Challenge 4

Answer the following questions.

**1  a**  The ratio of the lengths of the short sides of a right-angled triangle is 2:1. Calculate the size of the smallest angle in the triangle.

_____

_____

_____

**b**  Calculate the ratio of the length of the hypotenuse to the length of the shortest side of the triangle. Write your answer in the form $x$:1.

_____

_____

_____

**2  a**  The ratio of the length of the hypotenuse to the length of the shortest side of a right-angled triangle is 7:2. Calculate the size of the smallest angle in the triangle.

_____

_____

_____

**b**  If the hypotenuse is 21 cm long, calculate the lengths of the remaining two sides.

_____

_____

_____

**3**  Write an algebraic expression for length $y$.

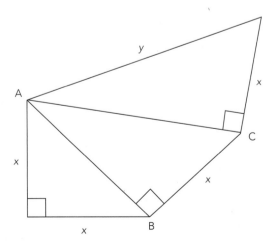

_____

_____

_____

_____

_____

_____

ISBN: 9780170477680    PHOTOCOPYING OF THIS PAGE IS RESTRICTED UNDER LAW.

**4**

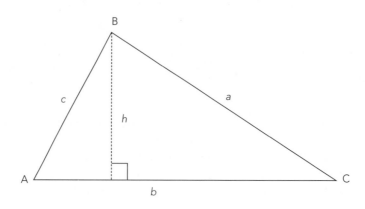

**a** Write an expression for the vertical height, *h*, in terms of angle BCA and side length *a*.

_____

_____

**b** Substitute this relationship into the traditional formula for the area of a triangle ($A = \dfrac{1}{2}$ x base x height). This will produce a new formula for finding the area of a triangle.

_____

_____

**c** Describe the differences between the information you would need about a triangle in order to use each formula.

_____

_____

**5** Triangle ABC is equilateral with sides 2 units long. BD bisects angle ABC and is a perpendicular bisector of the line AC.

**a** Show that the length of BD is $\sqrt{3}$.

_____

_____

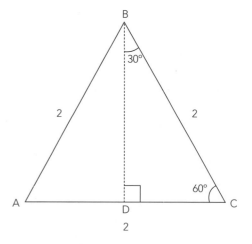

**b** Using the lengths 1, 2 and $\sqrt{3}$, complete the table.

| | Value | | Value |
|---|---|---|---|
| sin 60° | $\dfrac{\sqrt{3}}{2}$ | sin 30° | |
| cos 60° | | cos 30° | |
| tan 60° | | tan 30° | |

PHOTOCOPYING OF THIS PAGE IS RESTRICTED UNDER LAW.
ISBN: 9780170477680

# Pythagoras and trigonometry in 3D

- When asked to find a length or angle within a 3D shape, it is a good idea to draw the 2D shape that contains the length or angle.
- These usually involve right-angled triangles, so you can expect to use the Theorem of Pythagoras and trigonometry.

**Example:** This figure represents a cuboid.

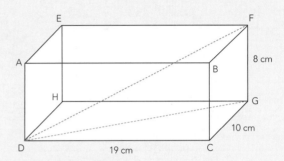

**a** Calculate the length of DG.
DG lies in a triangle on the 'floor' of the cuboid.

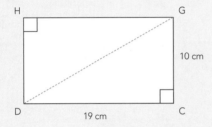

$$DG^2 = 19^2 + 10^2 \text{ (Pythagoras)}$$
$$\therefore DG = 21.47 \text{ cm (2 dp)}$$

**b** Calculate the length of DF. DF lies on the triangle DFG.

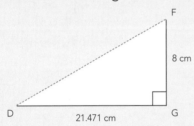

$$DF^2 = 21.471^2 + 8^2 \text{ (Pythagoras)}$$
$$\therefore DF = 22.91 \text{ cm (2 dp)}$$

**c** Calculate the angle between the line DF and the plane HGCD. ∠FDG lies on triangle DFG.

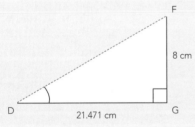

$$\angle FDG = \tan^{-1}\left(\frac{8}{21.471}\right)$$
$$\therefore \angle FDG = 20.4° \text{ (1 dp)}$$

Hint: Where possible, use the given values, not values you have calculated.

**d** Calculate the angle between the plane HFD and the plane DHGC:
∠FHG lies on triangle FGH.

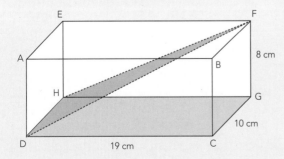

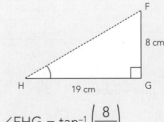

$$\angle FHG = \tan^{-1}\left(\frac{8}{19}\right)$$
$$\therefore \angle FHG = 22.8°$$

Answer the following questions.

**1** This figure represents a cuboid.

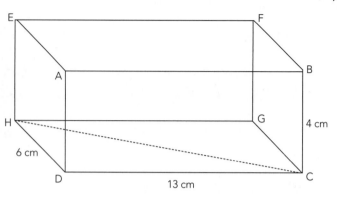

**a** Calculate the length of HC.

_____

_____

_____

**b** Calculate the length of HB.

_____

_____

_____

**c** Calculate the angle between HB
and the plane HGCD.

_____

_____

_____

**d** Calculate the angle between the
plane HBF and the plane HGCD.

_____

_____

_____

**e** Calculate the angle between HB
and the plane EADH.

_____

_____

_____

**f** Calculate the angle between HB
and the plane ABCD.

_____

_____

_____

PHOTOCOPYING OF THIS PAGE IS RESTRICTED UNDER LAW.     ISBN: 9780170477680

**2** This figure represents a cuboid.

**a** Calculate the length of GC.

_____

_____

_____

_____

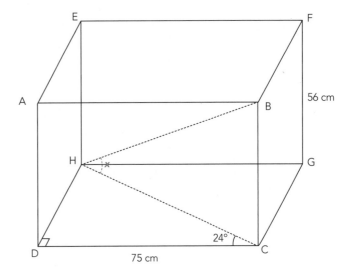

56 cm

75 cm

24°

**b** Calculate the angle x between HB
and the plane HGCD.

_____

_____

_____

_____

**c** Calculate the angle between the
plane HBF and the plane ABFE.

_____

_____

_____

_____

**3** This is a right cone.

**a** Calculate its slant height (AC).

_____

_____

_____

_____

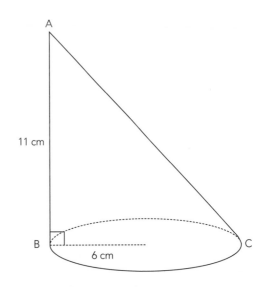

11 cm

6 cm

**b** Calculate the angle between AC
and its base.

_____

_____

_____

_____

**4** **a** Calculate the vertical height of the pyramid.

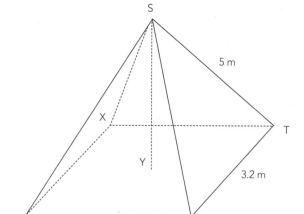

_____

_____

_____

**b** Calculate the angle between SU and the floor of the pyramid.

_____

_____

_____

**c** Calculate the angle between the plane TSU and the floor of the pyramid.

_____

_____

_____

**5** This figure is a triangular prism.

**a** Calculate the length of DC.

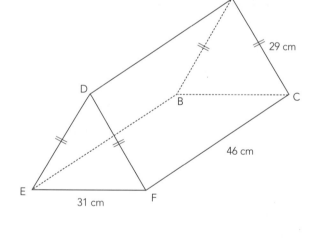

_____

_____

_____

**b** If G is the midpoint of EF, find the distance between G and A.

_____

_____

_____

**c** Calculate the angle between planes DACF and BCFE.

_____

_____

_____

PHOTOCOPYING OF THIS PAGE IS RESTRICTED UNDER LAW.     ISBN: 9780170477680

# Similar triangles

- The properties of similar shapes are described below for triangles.
- Similar shapes are **enlargements** of each other.

- Similar triangles are the same shape.
- They have the same-sized angles as each other.

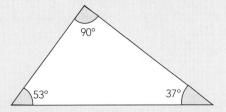 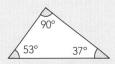

- However, they are not always drawn the same way up.

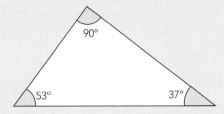 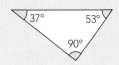

- The angles can also be in reverse order.

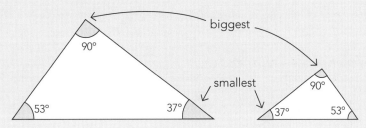

- The sides are also in proportion.

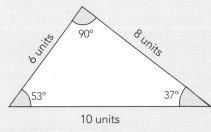 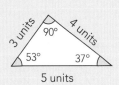

In this case: $\dfrac{\text{big triangle}}{\text{small triangle}} = \dfrac{10}{5} = \dfrac{8}{4} = \dfrac{6}{3} = 2$ — This is called **scale factor**.

- It is customary to label the vertices in corresponding order.

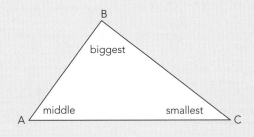 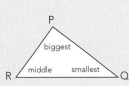

In this case: Δ**BAC** is similar to Δ**PRQ**.

Both are in **biggest, middle, smallest** order.

ISBN: 9780170477680     PHOTOCOPYING OF THIS PAGE IS RESTRICTED UNDER LAW.

## Justification of similar triangles using angles

- To show two triangles are similar, you need to show that two pairs of corresponding angles are equal.
- You need to **justify** each step in your reasoning.

**Examples:**

**1**

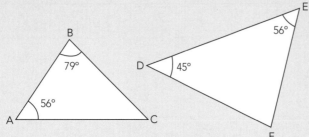

∠C = 45° (∠s in Δ = 180°)

∴ Two angles in each triangle are equal.

∴ ΔABC is similar to ΔEFD.

> Notice that the letters are written in the same angle order.
> ∠A = ∠E = 56°
> ∠B = ∠F = 79°
> ∠C = ∠D = 45°

**2**

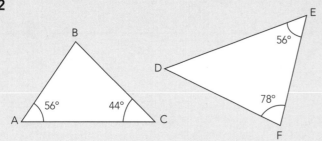

∠B = 80° (∠s in Δ = 180°)

∴ There are **not** two equal angles in each triangle.

∴ ΔABC is **not** similar to ΔEFD.

Show whether these shapes are similar or not.

**1**

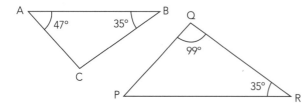

_____

_____

_____

_____

**2**

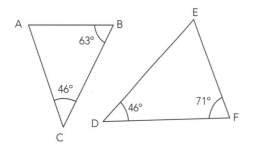

_____

_____

_____

_____

_____

**3**

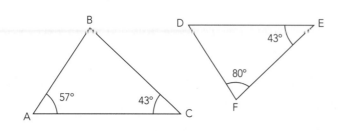

_____

_____

_____

_____

PHOTOCOPYING OF THIS PAGE IS RESTRICTED UNDER LAW.    ISBN: 9780170477680

## Justification of similar triangles using sides

- You need to show that **all** pairs of equivalent sides are proportional.
- The ratio of the sides is the scale factor of the enlargement.

1.25 is the scale factor for enlarging ABC to PRQ.

**Examples:**

**1**

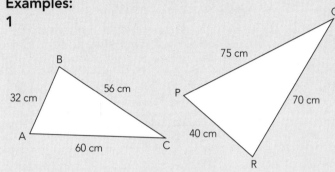

Shortest sides: $\dfrac{PR}{AB} = \dfrac{40}{32} = 1.25$

$\dfrac{RQ}{BC} = \dfrac{70}{56} = 1.25$

Longest sides: $\dfrac{QP}{CA} = \dfrac{75}{60} = 1.25$

∴ ΔABC and ΔPRQ are similar because all their sides are proportional.

**2**

B
7.2 cm
9.6 cm
A
8.4 cm
C
6.4 cm
P
5.6 cm
Q
5.0 cm
R

Longest sides: $\dfrac{BC}{PQ} = \dfrac{9.6}{6.4} = 1.5$

Shortest sides: $\dfrac{AB}{RQ} = \dfrac{7.2}{5.0} = 1.44$

∴ ΔABC and ΔPQR are not similar because two pairs of sides are not proportional.

69

Show whether these shapes are similar or not and calculate the scale factor.

**1**

_____

_____

_____

Scale factor = _____

**2**

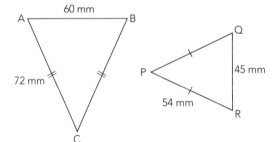

_____

_____

_____

Scale factor = _____

**3**

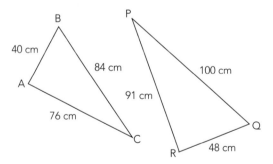

_____

_____

_____

Scale factor = _____

ISBN: 9780170477680 PHOTOCOPYING OF THIS PAGE IS RESTRICTED UNDER LAW.

## Calculating lengths of unknown sides

• Because the sides of similar triangles are in proportion, if we know the lengths of a pair of corresponding sides, we can use their proportion to calculate the lengths of other sides.

**Example:**

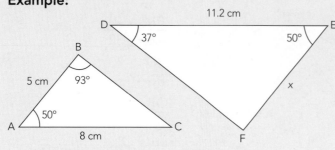

$\triangle ABC$ is similar to $\triangle EFG$.

So: $\dfrac{AC}{ED} = \dfrac{AB}{x}$   or   $\dfrac{ED}{AC} = \dfrac{11.2}{8}$

$\dfrac{8}{11.2} = \dfrac{5}{x}$   $= 1.4$

$x = 5 \times 1.4$

$x = \dfrac{11.2 \times 5}{8}$   $= 7$ cm

$x = 7$ cm

Find the missing lengths for these similar shapes.

**4**

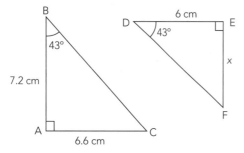

_____

_____

_____

_____

_____

**5**

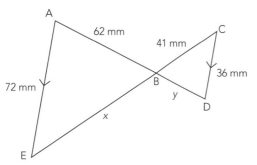

_____

_____

_____

_____

_____

**6**

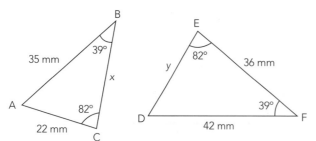

_____

_____

_____

_____

_____

**7**

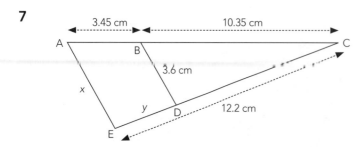

_____

_____

_____

_____

PHOTOCOPYING OF THIS PAGE IS RESTRICTED UNDER LAW.   ISBN: 9780170477680

**8**

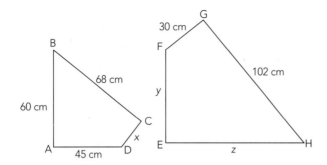

Answer the following questions.

**9** Calculate the lengths of AD and FC.

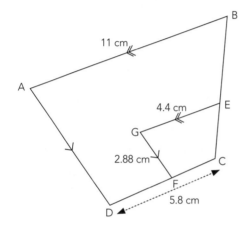

**10** Lines AB and ED are parallel. Calculate the length of AE.

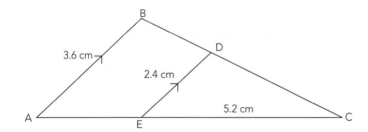

**11** Rectangles ACDF and FABE are similar. Calculate the area of rectangle ABEF.

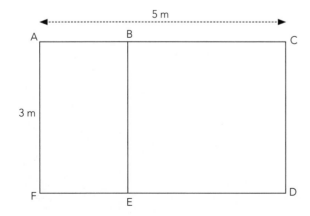

# Measurement

## Three-dimensional shapes: surface area and volume

Complete the table to make your own list of formulae.

| Shape | Volume | Surface area |
|---|---|---|
| Triangular prism | | |
| Cylinder | | |
| Cone | | |
| Pyramid | | |
| Sphere | | |

PHOTOCOPYING OF THIS PAGE IS RESTRICTED UNDER LAW.     ISBN: 9780170477680

# Surface area

- The surface area of an object is the sum of the areas of all its faces and surfaces.
- Drawing the net of the shape can be helpful.
- Don't forget to include square units in your answer.

**Examples:**

## Cylinder

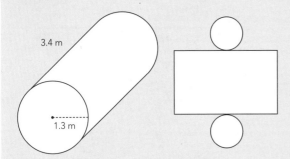

Surface area = area of curved side + 2(area of end)

$$= (3.4 \times \pi \times 2.6) + 2(\pi \times 1.3^2)$$

$$= 38.39 \text{ m}^2 \,(2 \text{ dp})$$

## Triangular prism

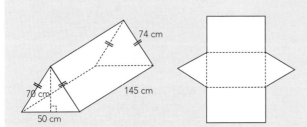

Surface area = 2(area side) + area base + 2(triangular end)

$$= 2(145 \times 74) + (50 \times 145) + 2(\tfrac{1}{2} \times 70 \times 50)$$

$$= 32\ 210 \text{ cm}^2$$

## Pyramid

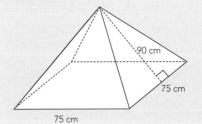 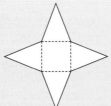

Surface area = area base + 4(area side)

$$= (75 \times 75) + 4(\tfrac{1}{2} \times 75 \times 90)$$

$$= 19\ 125 \text{ cm}^2$$

## Sphere

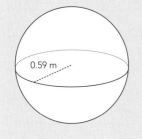

Surface area = $4\pi r^2$

$$= 4 \times \pi \times 0.59^2$$

$$= 4.37 \text{ m}^2 \,(2 \text{ dp})$$

## Cone

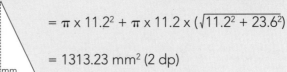

Surface area = $\pi r^2 + \pi r l$

*l* is the slant height of the cone, which may or may not be given.

$$= \pi \times 11.2^2 + \pi \times 11.2 \times (\sqrt{11.2^2 + 23.6^2})$$

$$= 1313.23 \text{ mm}^2 \,(2 \text{ dp})$$

Calculate the surface areas of the following shapes.

**1**

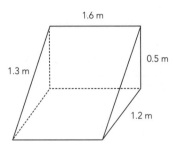

Net

_____

_____

_____

_____

**2**

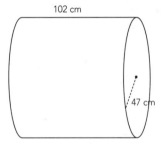

Net

_____

_____

_____

_____

**3**

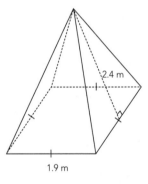

Net

_____

_____

_____

_____

**4**

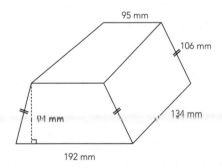

Net

_____

_____

_____

_____

PHOTOCOPYING OF THIS PAGE IS RESTRICTED UNDER LAW.   ISBN: 9780170477680

**5**

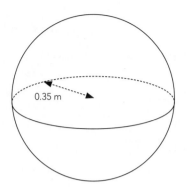

_____

_____

_____

_____

**6**

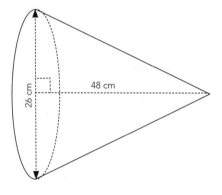

_____

_____

_____

_____

**7**

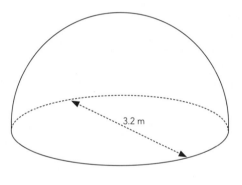

_____

_____

_____

_____

**8**

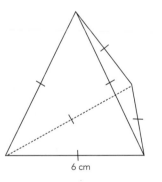

_____

_____

_____

_____

**9**   The diagram represents a solid cylinder with a hemisphere hollowed out of the top.

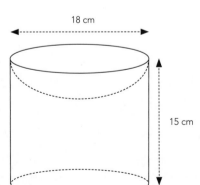

_____

_____

_____

_____

# Volume

- The volume of a three-dimensional (3D) shape is the amount of space the shape occupies.
- Don't forget to add cubic units to your answer.

## Prisms and cylinders

**Volume = area of face x depth**

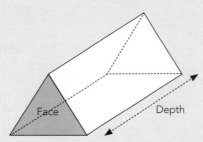

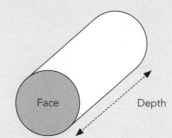

Calculate the volumes of the following shapes.

**1**

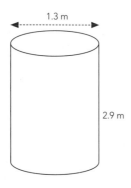

1.3 m
2.9 m

_____

_____

_____

**2**

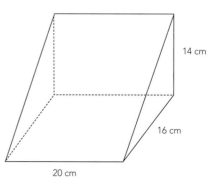

14 cm
16 cm
20 cm

_____

_____

_____

**3**

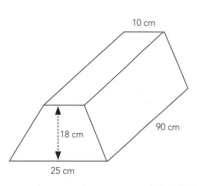

10 cm
18 cm
90 cm
25 cm

_____

_____

_____

**4**

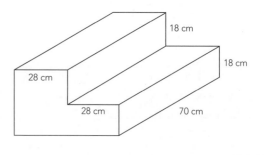

18 cm
18 cm
28 cm
28 cm
70 cm

_____

_____

_____

PHOTOCOPYING OF THIS PAGE IS RESTRICTED UNDER LAW.   ISBN: 9780170477680

## Pyramids, cones and spheres

### Pyramid
- A pyramid can have any polygon as its base, and its sides are triangles.

$$\text{Volume of pyramid} = \frac{1}{3} \times \text{(base area)} \times \text{height}$$

**Example:**

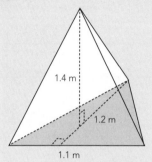

You must use the **vertical** height.

$$\text{Volume} = \frac{1}{3} \times \text{(base area)} \times \text{vertical height}$$

$$= \frac{1}{3} \times (\frac{1}{2} \times 1.1 \times 1.2) \times 1.4$$

$$= 0.308 \text{ m}^3 \text{ (1 dp)}$$

### Cone
- A cone is similar to a pyramid, but it has a circular base and the sides are curved.

$$\text{Volume of cone} = \frac{1}{3} \times \text{(base area)} \times \text{height}$$

$$= \frac{1}{3} \pi r^2 h$$

The base is always a circle.

**Example:**

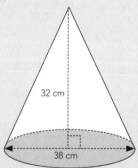

$$\text{Volume} = \frac{1}{3} \times \text{(base area)} \times \text{vertical height}$$

$$= \frac{1}{3} \times (\pi \times 19^2) \times 32$$

$$= 12\ 097.23 \text{ m}^3 \text{ (2 dp)}$$

Remember to use the radius, not the diameter.

### Sphere
- A basketball is the shape of a sphere.

$$\text{Volume of sphere} = \frac{4}{3} \pi r^3$$

**Example:**

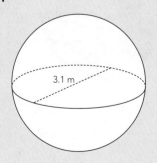

$$\text{Volume} = \frac{4}{3} \times \pi \times 1.55^3$$

$$= 15.6 \text{ m}^3 \text{ (1 dp)}$$

Remember to round appropriately.

Calculate the volumes of these shapes.

**5**

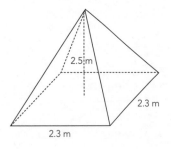

2.5 m
2.3 m
2.3 m

_____

_____

**6**

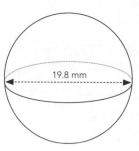

19.8 mm

_____

_____

**7**

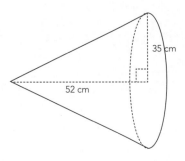

35 cm
52 cm

_____

_____

**8**

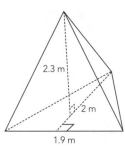

2.3 m
2 m
1.9 m

_____

_____

**9**

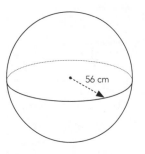

56 cm

_____

_____

**10**

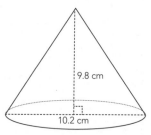

9.8 cm
10.2 cm

_____

_____

**11**

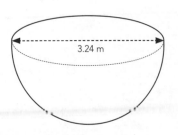

3.24 m

_____

_____

**12**

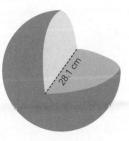

28.1 cm

_____

_____

PHOTOCOPYING OF THIS PAGE IS RESTRICTED UNDER LAW.    ISBN: 9780170477680

## Areas and volumes with ratios

Consider a series of lines, squares and cubes:

| Length | Area | Volume |
|---|---|---|
| │<br>1 unit | ☐<br>$1^2$ or 1 unit$^2$ | $1^3$ or 1 unit$^3$ |
| ┼<br>2 units | $2^2$ or 4 units$^2$ | $2^3$ or 8 units$^3$ |
| ╪<br>4 units | $4^2$ or 16 units$^2$ | $4^3$ or 64 units$^3$ |
| Length | Length$^2$ | Length$^3$ |

**Compare the first and second lines:**

Ratio of original length:new length = 1 : 2

⇒ original area:new area = $1^2 : 2^2$ or 1 : 4 and original volume:new volume = $1^3 : 2^3$ or 1 : 8.

**Compare the first and third lines:**

Ratio of original length:new length = 1 : 4

⇒ original area:new area = $1^2 : 4^2$ or 1 : 16 and original volume:new volume = 1 : $4^3$ or 1 : 64.

**In general:**

Ratio of original length : new length = 1 : $x$

⇒ original area : new area = 1 : $x^2$

and original volume : new volume = 1 : $x^3$

**Examples:**

1   The ratio of the dimensions of a smaller cube to those of a larger cube is 1:5. If the volume of the smaller cube is 100 cm$^3$, what is the volume of the larger cube?

Volume larger cube = Volume smaller cube x $5^3$

= 100 x 125

= 12 500 cm$^3$

2   The ratio of the dimensions of a larger rectangle to those of a smaller rectangle is 5:3. If the area of the larger rectangle is 76 cm$^2$, what is the area of the smaller rectangle?

5 : 3 = 1 : 0.6          Area smaller rectangle = Area larger rectangle x $(0.6)^2$

= 76 x $(0.6)^2$

= 27.36 cm$^2$

Divide both sides by 5.

**1**  The ratio of the dimensions of a smaller cube to those of a larger cube is 1:3. If the smaller cube holds 1 L, what is the capacity of the larger cube?

**2**  The ratio of the dimensions of a larger cylinder to those of a smaller cylinder is 2:1. If the larger cylinder holds 480 cm³, what is the capacity of the smaller cylinder?

**3**  The ratio of the dimensions of a smaller triangle to those of a larger triangle is 1:2.5. If the area of the smaller triangle is 10 cm², calculate the area of the larger triangle.

**4**  The ratio of the dimensions of a larger pyramid to those of a smaller similar pyramid is 4:3. If the volume of the larger pyramid is 240 m³, calculate the volume of the smaller pyramid.

**5**  The ratio of the dimensions of a larger circle to those of a smaller circle is 5:2. If the area of the larger circle is 195 m², calculate the area of the smaller circle.

**6**  The ratio of the dimensions of an enlarged 2D shape to those of the original shape is 3:1. If the area of the original shape is 5xy, write an expression for the area of the new shape.

**7**  The radius of a sphere is increased by half. If it originally had a volume of 1 L, what will the new volume be?

**8**  The volume of a figure increases from 3 L to 192 L. Calculate the ratio of the increase in its dimensions.

**9**  The area of a figure decreases from 1675 cm² to 67 cm². Calculate the ratio of the decrease in its dimensions.

**10**  The volume of a figure increases from 50 L to 137.2 L. Calculate the ratio of the increase in its dimensions.

PHOTOCOPYING OF THIS PAGE IS RESTRICTED UNDER LAW.    ISBN: 9780170477680

## Compound shapes: surface area and volume

- Compound shapes are made up of other simple shapes.
- You will need to find the area or volume of the separate shapes and combine them.
- Sometimes there are several ways of calculating areas and volumes.
- Do not round until the end of your calculations.
- If you have to round intermediate values, round them to at least one more decimal place than the information you have been given.

**Example:**

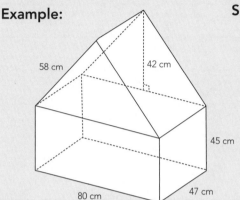

**Surface area:**

| | |
|---|---|
| Area of side walls | $= 2(80 \times 45 + 47 \times 45)$ |
| | $= 11\ 430\ \text{cm}^2$ |
| Area of end triangles | $= 2(\frac{1}{2} \times 80 \times 42)$ |
| | $= 3360\ \text{cm}^2$ |
| Area of roof | $= 2(58 \times 47)$ |
| | $= 5452\ \text{cm}^2$ |
| Area of bottom | $= 80 \times 47$ |
| | $= 3760\ \text{cm}^2$ |
| Total area | $= 24\ 002\ \text{cm}^2$ |

**Volume:**

Volume of triangular prism $= \frac{1}{2} \times 80 \times 42 \times 47$

$= 78\ 960\ \text{cm}^3$

Volume of cuboid $= 80 \times 47 \times 45$

$= 169\ 200\ \text{cm}^3$

Total volume $= 248\ 160\ \text{cm}^3$

Calculate the total surface area and volume of each shape.

**1** This shape is made up of a cylinder and a cone.

2.1 m

2.3 m

0.8 m

_____

_____

_____

_____

_____

_____

_____

_____

_____

_____

ISBN: 9780170477680    PHOTOCOPYING OF THIS PAGE IS RESTRICTED UNDER LAW.

**2** This shape is made up of a cuboid and a hemisphere with a radius of 0.28 m.

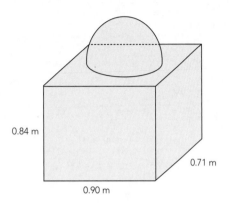

0.84 m

0.71 m

0.90 m

_____
_____
_____
_____
_____
_____
_____
_____
_____
_____
_____

**3** This shape is made up of a cuboid with an equilateral triangular prism taken from inside it. The height of the equilateral triangle is 16.5 cm.

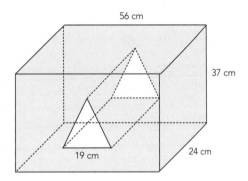

56 cm

37 cm

19 cm

24 cm

_____
_____
_____
_____
_____
_____
_____
_____
_____
_____

**4** This shape is like a piece of cylindrical pipe that is 24 cm long.

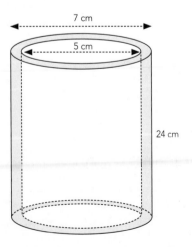

7 cm

5 cm

24 cm

_____
_____
_____
_____
_____
_____
_____
_____

PHOTOCOPYING OF THIS PAGE IS RESTRICTED UNDER LAW.    ISBN: 9780170477680

**5** This shape is made up of a cuboid with a half cylinder removed from it.

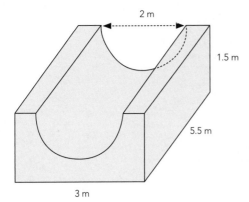

**6** This shape is a cylinder with a cone removed from the inside.

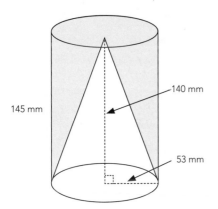

**7** This shape is part of a cone and is called a frustum. The height of the frustum (1.3 m) is two thirds that of the cone.

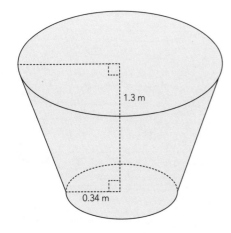

ISBN: 9780170477680    PHOTOCOPYING OF THIS PAGE IS RESTRICTED UNDER LAW.

## Working backwards

- Sometimes, you can use the formula for an area or volume to calculate an unknown dimension.

**Examples:**

**1** The volume of a cone is 0.8244 m³. If its height is 1.23 m, calculate the radius.

$$\text{Volume of cone} = \frac{1}{3} \times \pi r^2 \times h$$

$$\frac{1}{3} \times \pi r^2 \times 1.23 = 0.8244$$

$$r^2 = \frac{0.8244}{0.41\pi}$$

Isolate the unknown dimension on one side.

$$r = \sqrt{0.64}$$

$$\text{Radius} = 0.8 \text{ m}$$

**2** The surface area of this square-based pyramid is 96 cm². Calculate its base length.

$$\text{Surface area of a pyramid} = 4\left(\frac{1}{2} \times b \times h\right) + b^2$$

$$4\left(\frac{1}{2} \times b \times 5\right) + b^2 = 96$$

$$10b + b^2 = 96$$

$$b^2 + 10b - 96 = 0$$

$$(b + 16)(b - 6) = 0$$

$$b = -16 \text{ or } 6$$

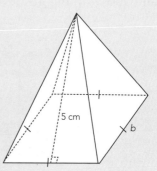

The base can't be a negative length, so it is 6 cm long.

Answer the following questions.

**1** This cylinder has a volume of 9.31 cm³. Calculate the diameter.

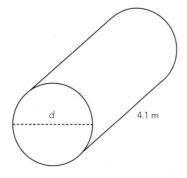

**2** The surface area of a sphere is 14 103 cm². Calculate its radius.

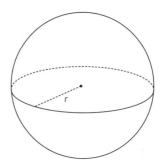

_____

_____

_____

_____

PHOTOCOPYING OF THIS PAGE IS RESTRICTED UNDER LAW.   ISBN: 9780170477680

**3** The surface area of this regular tetrahedron is 48 m². Calculate the length of each edge.

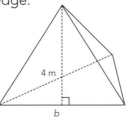

_____

_____

_____

_____

**4** The total surface area of this hemisphere is 235.6 cm². Calculate its diameter.

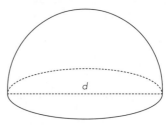

_____

_____

_____

**5** The walls of this section of pipe are 2 cm thick. The volume of material in the walls is 4800π. Calculate its internal radius (r).

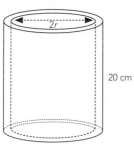

_____

_____

_____

**6** This shape is a cuboid with a square-based pyramid on top. The volume is 490 m³. Calculate the length of the base.

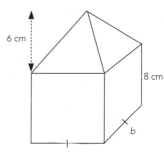

_____

_____

_____

**7** The volume of this square-based frustum is 241 m³. The full height of the pyramid would be 12 cm. Calculate the length of the base.

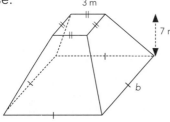

_____

_____

_____

_____

**8** The surface area of this cone is 48π cm². Calculate the radius.

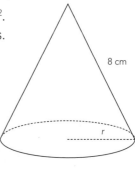

_____

_____

_____

_____

# Challenge 5

Answer the following questions.

**1** Calculate the surface area of this triangular prism.

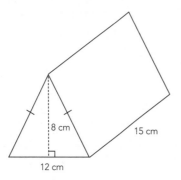

_____

_____

_____

_____

_____

**2** Create an expression for the surface area of this square-based pyramid.

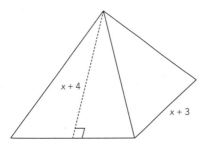

_____

_____

_____

_____

_____

**3** Show that the surface area of this cuboid is SA = $6y^2 + 44y + 56$.

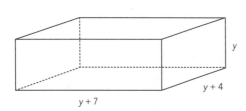

_____

_____

_____

_____

_____

**4** This square-based pyramid has sides that are equilateral triangles. Calculate the volume of this pyramid.

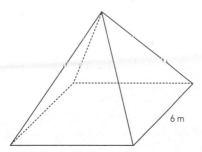

_____

_____

_____

_____

_____

 PHOTOCOPYING OF THIS PAGE IS RESTRICTED UNDER LAW. ISBN: 9780170477680

**5**    Given that the surface area of this cone is $28\pi$, show that $r^2 + 3r - 28 = 0$, and find its radius.

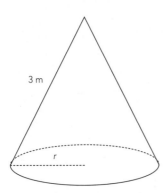

_____

_____

_____

_____

_____

_____

**6**    The surface area of this cube and sphere are equal. Find the value of $a$ when $y = \sqrt{a\pi}$.

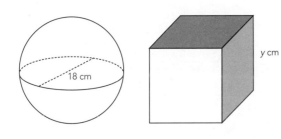

_____

_____

_____

_____

_____

_____

**7**    This shape is made up of a hemisphere and cone. The ratio of the height of the cone to the height of the hemisphere is 3:5. Calculate the volume of this compound shape.

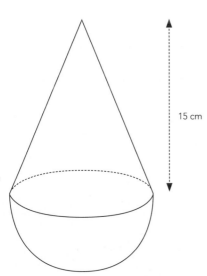

_____

_____

_____

_____

_____

_____

_____

_____

**8**   Two spheres with a diameter of 5 cm fit inside a cylinder.
Calculate the percentage of the cylinder that is filled by the spheres.

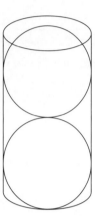

_____

_____

_____

_____

_____

_____

_____

**9**   The volume of the hemisphere is twice the volume of the cone. Find an expression for _h_ in terms of _y_.

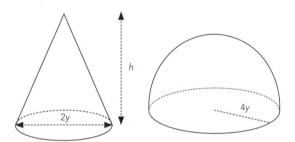

_____

_____

_____

_____

_____

_____

_____

**10**   The volume of a cylinder is given by $V = \pi r^2 h$ and the volume of a cone is given by $V = \frac{1}{3}\pi r^2 h$.

A cylinder has the same height as a cone. If the volume of the cylinder is six times the volume of the cone, write an expression for the ratio of the radius of the cylinder to the radius of the cone.

_____

_____

_____

_____

_____

_____

_____

_____

PHOTOCOPYING OF THIS PAGE IS RESTRICTED UNDER LAW.   ISBN: 9780170477680

 Proof

- A proof is a mathematical argument.
- You need to justify each step in a proof.
- Proofs must be general, so they use letters rather than numbers.
- Frequently, a previous question will hint at how a proof can be done.
- It is usually easiest to start with the more complex side and show that this is equal to the simpler side.
- Sometimes there are several different ways of proving a mathematical statement.

## Geometric proof

- Sometimes, devising a proof requires the construction of extra lines.
- It can also be useful to name angles $x$, $y$, etc.

**Examples:**

1  Prove that the exterior angle of a triangle = the sum of the internal opposite angles ($\angle DBA = \angle BAC + \angle ACB$).

Construct a line through B that is parallel to AC. Label it BE.

$\angle DBE = \angle BCA$ (corr $\angle$s =, // lines)
$\angle EBA = \angle BAC$ (alt $\angle$s =, // lines)
$\angle DBA = \angle DBE + \angle EBA$
$\therefore \angle DBA = \angle BCA + \angle BAC$

2  Prove that the diagonals of a rhombus intersect at right angles.

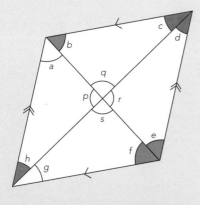

$\angle b = \angle e$ (isos $\triangle$, base $\angle$s =)
$\angle b = \angle f$ (alt $\angle$s =, // lines)
$\therefore \angle e = \angle f$

$\angle h = \angle c$ (isos $\triangle$, base $\angle$s =)
$\angle h = \angle d$ (alt $\angle$s =, // lines)
$\therefore \angle c = \angle d$

$(c + d) + (e + f) = 180°$ (co-int $\angle$s add to 180°, // lines)
$\therefore d + e = 90°$
$\therefore r = 90°$ ($\angle$s in $\triangle \Rightarrow 180°$)

You do not need to repeat your argument — use the word 'similarly'. ⟶ Similarly, $s$, $p$ and $q$ equal 90°.

ISBN: 9780170477680    PHOTOCOPYING OF THIS PAGE IS RESTRICTED UNDER LAW.

Fill in the gaps to complete the following proofs.

**1**  AB is parallel with DE.
Prove that $\angle r = \angle q + \angle t$.

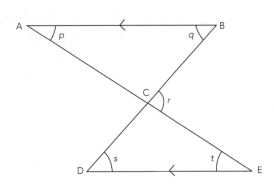

Step 1: Construct a line parallel to AB and ED that passes through C. Call this FG.

Step 2: Use parallel line rules:

$$\angle q = \angle BCG \ (\ \underline{\hspace{4cm}}\ )$$

$$\angle t = \angle\underline{\hspace{1cm}} \ (\ \underline{\hspace{4cm}}\ )$$

$$\angle r = \angle BCG + \angle\underline{\hspace{2cm}}$$

$$\therefore \angle r = \angle q + \angle t$$

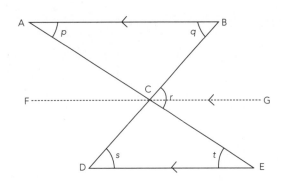

**2**  DE is parallel with BC. Prove that triangles ABC and ADE are similar.

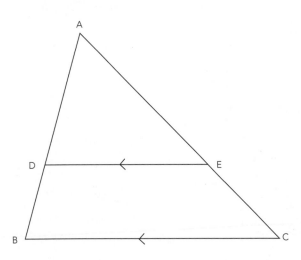

$$\angle ADE = \angle\underline{\hspace{1cm}} \ (\ \underline{\hspace{4cm}}\ )$$

$$\angle\underline{\hspace{1cm}} = \angle\underline{\hspace{1cm}} \ (\ \underline{\hspace{4cm}}\ )$$

$\angle A$ is common to both triangles

$\therefore$ triangles ABC and ADE have $\underline{\hspace{2cm}}$

$\underline{\hspace{5cm}}$

$\therefore$ triangles ABC and ADE are $\underline{\hspace{3cm}}$

PHOTOCOPYING OF THIS PAGE IS RESTRICTED UNDER LAW.    ISBN: 9780170477680

Complete the following proofs.

**3** DE is parallel with AC. Use this diagram to prove that the sum of the angles in any triangle is 180°.

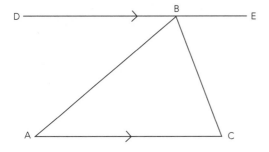

_____
_____
_____
_____
_____
_____
_____
_____
_____
_____
_____
_____

**4** AC is parallel with DG. Angle CBE is bisected by BF. Show that triangle BEF is isosceles.

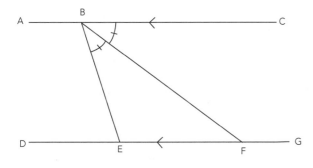

_____
_____
_____
_____
_____
_____
_____
_____
_____
_____
_____
_____

# Algebraic proof

- An algebraic proof is a mathematical argument that usually does not use numbers.
- Often the question will ask you to 'Show that …'.

**Hints:**
- Even numbers: call these $2x$, e.g. 10, where $x = 5$.
- Odd numbers: call these $2x + 1$, e.g. 15, where $x = 7$.
- Consecutive numbers come after each other: call these $x$, $x + 1$, $x + 2$, e.g. 6, 7, 8, where $x = 6$.
- Consecutive even numbers: call these $2x$, $2x + 2$, $2x + 4$, e.g. 6, 8, 10, where $x = 3$.
- Consecutive odd numbers: call these $2x + 1$, $2x + 3$, $2x + 5$, e.g. 9, 11, 13, where $x = 4$.
- A prime number can only be divided by two numbers: 1 and itself $\therefore$ 1 is not a prime number.

**Examples:**

**1** Show that for any integer ($n$) other than 1, $n^2 + n$ can never be prime.

Let $n$ be even, so $n = 2x$:

$$(2x)^2 + (2x) = 4x^2 + 2x$$
$$= 2(2x^2 + x)$$

As a multiple of two, this cannot be prime.

Let $n$ be odd, so $n = 2x + 1$:

$$(2x + 1)^2 + (2x + 1) = 4x^2 + 4x + 1 + 2x + 1$$
$$= 4x^2 + 6x + 2$$
$$= 2(2x^2 + 3x + 1)$$

As a multiple of two, this cannot be prime.

$\therefore n^2 + n$ can never be prime.

Alternative proof:

$n^2 + n = n(n + 1)$, which is a product of 2 integers. A product can never be a prime.

**2** Use algebra to show that any whole number under 100 that ends with 5 has a square that ends with 25.

Let the number be written '$a5$'.

It has the value $(a \times 10 + 5) = 10a + 5$

$$('a5')^2 = (10a + 5)^2$$
$$= 100a^2 + 100a + 25.$$

Both $100a^2$ and $100a$ are multiples of 100, so their sum will also be a multiple of 100 and end with two zeros.

$\therefore$ Any square of a whole number under 100 and ending in 5 must end in 25.

**3** Show that $(3n - 4)^2 + (n^2 + 1)$ is always odd.

$$(3n - 4)^2 + (n^2 + 1) = 9n^2 - 24n + 16 + n^2 + 1$$
$$= 10n^2 - 24n + 16 + 1$$
$$= 2(5n^2 - 12n + 8) + 1$$

$(5n^2 - 12n + 8)$ must be an integer.

2 x (any integer) must always be even.

$\therefore$ 2 x (any integer) + 1 must always be odd.

$\therefore (3n - 4)^2 + (n^2 + 1)$ is always odd.

PHOTOCOPYING OF THIS PAGE IS RESTRICTED UNDER LAW.     ISBN: 9780170477680

Answer the following.

**1** Show that the difference between two odd numbers is always even.

_____

_____

_____

_____

_____

**2** Prove that the difference between the squares of two consecutive integers equals the sum of these two integers.

_____

_____

_____

_____

_____

**3** Prove that for any integer $n$, $(3n + 1)^2 + (n - 1)^2$ is always even.

_____

_____

_____

_____

_____

**4** Prove that for any three consecutive integers, the difference between the squares of the first and last numbers is four times the middle number.

_____

_____

_____

_____

_____

**5**   For any integer $n$, prove that $(n + 3)(2n + 1) + (n - 2)(2n + 1)$ is not a multiple of 2.

_____

_____

_____

_____

_____

**6**   Prove that for any integer $n$, $(2n + 9)^2 - (2n + 5)^2$ is always a multiple of 4.

_____

_____

_____

_____

_____

**7**   Show that for any two-digit whole number ('$ab$') that is less than 100, if the digits add to 9, then the number ('$ab$') must be a multiple of 9.

_____

_____

_____

_____

_____

**8**   Prove that $(n + 1)^3 - (n + 1)^2$ is always a multiple of $n$.

_____

_____

_____

_____

_____

PHOTOCOPYING OF THIS PAGE IS RESTRICTED UNDER LAW.   ISBN: 9780170477680

# Practice sets

## Practice set one

**1** Simplify: $\dfrac{(6a^2)^3}{3a^4 \times 2a^3}$

**2** Expand and simplify: $-4x(2 - x + y) + 3x$

**3** Calculate the value of $A$ if $b = -2$, $c = 4$, $d = 7$ and $A = \dfrac{bd - dc}{c^2}$.

**4** If $A = \dfrac{4x - 3y}{x - 2y}$, give the equation for $x$ in terms of $y$ and $A$.

**5** Write an expression for the shaded area.

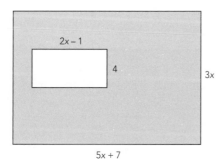

**6** Find the lengths of the three sides of this triangle.

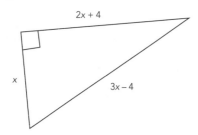

**7** Solve the inequation:

$2^{x^2} > 2^x \times 16^3$

_____

_____

_____

**8** Factorise completely:

$4x^2 + 36x + 56$

_____

_____

_____

**9** Write the equation for this graph.

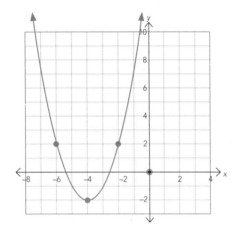

_____

_____

_____

**10** Write an expression for the perimeter of this shape in terms of x.

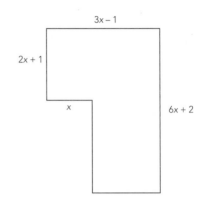

_____

_____

_____

**11** The volume of a hemisphere is $144\pi$. Calculate the total surface area.

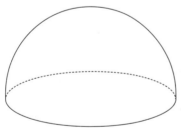

_____

_____

_____

_____

**12** Find the length of side CE.

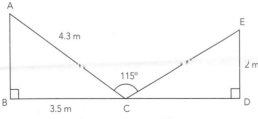

_____

_____

_____

_____

PHOTOCOPYING OF THIS PAGE IS RESTRICTED UNDER LAW.    ISBN: 9780170477680

# Practice set two

**1** Simplify this fraction as far as possible.

$$\frac{5x^2 + 25x}{x^2 - 25}$$

_____

_____

_____

**2** Solve for $x$ when $y = 5$ and $z = 60°$.

$$y^2 - 4x = 6\cos z$$

_____

_____

_____

**3** Expand and simplify:
$-x(5x + 2y - 3) + 4(-x + 6y - 7)$

_____

_____

_____

**4** A number has 6 added to it and it is then multiplied by 3. The result is 54. What is the number?

_____

_____

_____

**5** Find the ratio of the volume of the cuboid to the volume of the triangular prism. Give your answer in its simplest form.

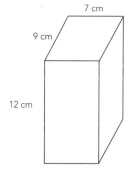
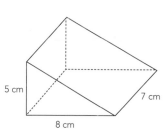

_____

_____

_____

_____

**6** The circumference of this cone is 18 π. Calculate the angle between BC and the circular top.

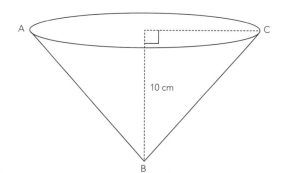

_____

_____

_____

_____

_____

**7** Solve the equation:
$2^{x^2 + 1} = 32 \div 8^x$

_____

_____

_____

**8** Solve for $x$ and $y$:
$8x + 10y = 54$
$5x - 4y = 44$

_____

_____

_____

**9** The perimeter of this shape is $15x + 10$. Find the length of side FG.

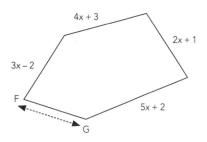

_____

_____

_____

**10** Write a quadratic equation that has a maximum at (4, 7) and a $y$ intercept of –9.

_____

_____

_____

_____

_____

_____

**11** Factorise: $5x^2 - 45$

_____

_____

_____

**12** Write as a simplified single fraction:

$$\frac{6x - 2}{3} + \frac{3x - 1}{4}$$

_____

_____

_____

**13** If 7 cm ≤ $x$ ≤ 10 cm and 5.1 cm ≤ $y$ ≤ 6.3 cm, complete this statement to show the range of values that the shaded area can take. Both figures are squares.

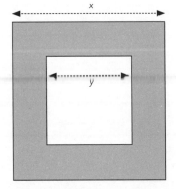

_____ ____ Area ____ _____

_____ ____ Area ____ _____

PHOTOCOPYING OF THIS PAGE IS RESTRICTED UNDER LAW.    ISBN: 9780170477680

# Practice set three

**1** Factorise this expression:
$27x^2 - 90x + 63$

_____

_____

_____

**2** If $x$ can take only integer values, solve the inequality $7 + 3x > 28 + x$.

_____

_____

_____

**3** Solve: $20 < \dfrac{x^2 + 4}{2} < 74$

_____

_____

_____

**4** If $x = -4$, $y = -3$ and $z = 7$, find the value of $6 + 2(3y^2 - x + z)$.

_____

_____

_____

**5** The volume of the cube is equal to 60% of the volume of the cuboid. Calculate the height of the cuboid.

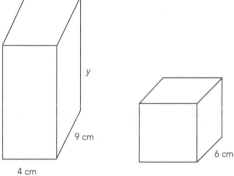

_____

_____

_____

_____

_____

_____

_____

**6** Calculate the area of each triangle.

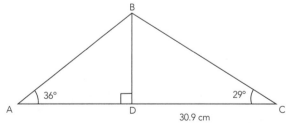

_____

_____

_____

_____

_____

_____

_____

**7** Simplify:

$$\frac{6a^2bc - 9ab^2}{15abc + 3a^2b - 12ab^2c^2}$$

_____

_____

_____

**8** Solve the inequality:

$(2x + 3)(2x - 1) \geq (x - 3)(4x + 2)$

_____

_____

_____

**9** Write the equation for this graph.

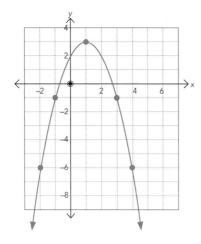

_____

_____

_____

_____

_____

_____

_____

_____

**10** This figure is made up of two overlapping figures. The ratios of the total area of rectangle to the shaded area to the total area of the circle are 5:1:3. The area of the whole shape is 147 m². Calculate the area of the rectangle.

_____

_____

_____

_____

_____

_____

_____

**11** Calculate the values of $x$ and $y$, and use them to find the dimensions of this rectangle.

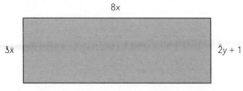

_____

_____

_____

_____

PHOTOCOPYING OF THIS PAGE IS RESTRICTED UNDER LAW.    ISBN: 9780170477680

# Practice set four

**1** Solve: $2^x = 256$

_____

_____

_____

**2** Solve the equation:

$$\frac{x + 3}{2} - \frac{2x - 1}{3} = 1$$

_____

_____

_____

**3** If $y = x^2 + 2x - 15$, for what values of $x$ will $y$ be negative?

_____

_____

_____

**4** The sides of a rectangle are $3x + 2$ and $x - 4$. Give an expression for the area of the rectangle in the form $ax^2 + bx + c$.

_____

_____

_____

**5** These rectangles are similar. What is the ratio of the larger perimeter to the smaller one?

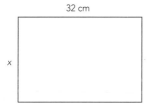

_____

_____

_____

_____

_____

_____

_____

**6** The curved surface area of this cone is $540\pi$ cm$^2$. The volume can be calculated in the form $a\pi$ cm$^3$. Calculate the value of $a$.

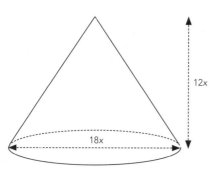

_____

_____

_____

_____

_____

_____

_____

**7** Solve: $2x^2 + 5 = 6x + x^2$

_____

_____

_____

**8** Solve: $2^{x^2 - 3x} > 2^{10}$

_____

_____

_____

**9** Show that a sphere of equal volume to this cylinder would have $r = \sqrt[3]{135}\,x$.

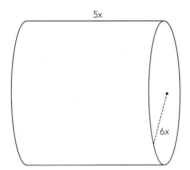

_____

_____

_____

_____

_____

_____

_____

**10** The perimeter of rectangle A is 44 cm and the perimeter of rectangle B is 28 cm.
Find the base and height of rectangles A and B.

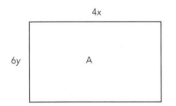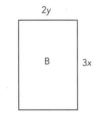

_____

_____

_____

_____

_____

_____

**11** The area of this triangle is 210 cm$^2$.
Calculate the value of $x$.

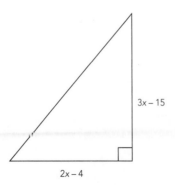

_____

_____

_____

_____

_____

_____

PHOTOCOPYING OF THIS PAGE IS RESTRICTED UNDER LAW.   ISBN: 9780170477680

## Practice set five

**1** Add these fractions and give your answer in its simplest form.

$$\frac{3x-1}{6} + \frac{x}{4}$$

_____

_____

_____

**2** Simplify:

$$\frac{x^2 - x - 12}{-27 + x^2 - 6x}$$

_____

_____

_____

**3** Fifty plus double a number is six more than seven times the number. If the answer is a prime number, what is the smallest value it can be?

_____

_____

_____

**4** What is the turning point of the equation $y = (x - 1)(x - 5)$?

Turning point: _____

This is a maximum/minimum.

**5** This cylinder has a surface area of $130\pi$ cm². Calculate the radius.

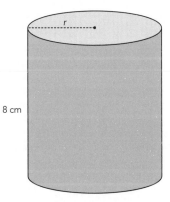

8 cm

_____

_____

_____

_____

_____

_____

_____

**6** Calculate the scale factor of these similar triangles and find the unknown lengths.

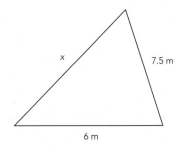

$x$     7.5 m

6 m

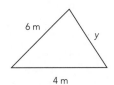

6 m     $y$

4 m

_____

_____

_____

_____

_____

_____

**7** Solve for $x$ and $y$:
$-3x + 7y = 1.125$
$-8x + 5y = 3$

_____

_____

_____

**8** Find a value for b so that $x^2 + bx + 36 = 0$ has exactly one solution.

_____

_____

_____

**9** Write the equation of this graph.

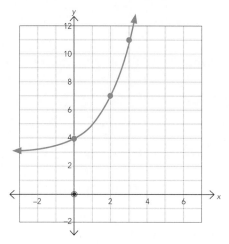

_____

**10** Give the equation for $b$ in terms of $a$, $c$, $d$ and $e$.
$ab^2 - 8c = 3d^2 + 12eb^2$

_____

_____

_____

**11** The surface area of the sphere is double the surface area of the cylinder. Calculate the ratio of the volume of the sphere to the volume of the cylinder.

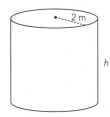

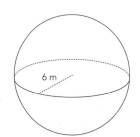

_____

_____

_____

_____

_____

_____

_____

**12** Find the volume of this prism.

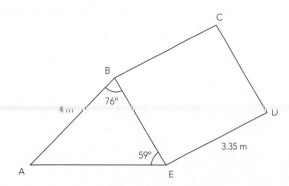

_____

_____

_____

_____

_____

_____

_____

PHOTOCOPYING OF THIS PAGE IS RESTRICTED UNDER LAW.   ISBN: 9780170477680

# Practice set six

**1**   Draw the graph of $y = 2^x - 3$.

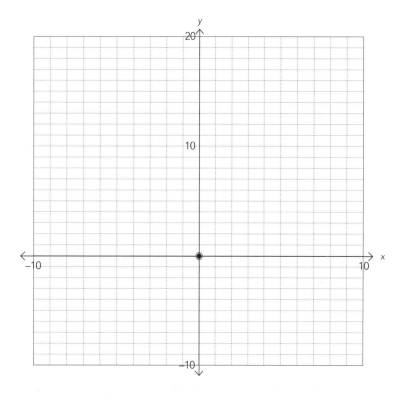

**2**   Calculate the equation of the line which connects (–4, 7) and (12, –5).

_____

_____

_____

_____

_____

**3**   Find the equation of a line which is perpendicular to $5y + 4x - 35 = 0$, and which passes through the point (20, –9).

_____

_____

_____

_____

_____

**4**   Write an equation for the quadratic sequence 0, 5, 16, 33, 56, …

_____

_____

_____

_____

_____

_____

**5** If $x = -1$, $y = -5$ and $z = 60°$, calculate the value of the expression $xy^2 - y\cos z$.

_____

_____

_____

_____

_____

**6** The diagonal of a rectangle is 17 cm long. The rectangle is 15 cm wide and $(x + 3)$ cm high. Calculate its area.

_____

_____

_____

_____

_____

**7** The ratio of the dimensions of a large cuboid to a smaller one is 7:2. If the volume of the smaller one is 248 cm³, calculate the volume of the larger cuboid.

_____

_____

_____

**8** Solve: $25 \times 5^{2n-1} = 125^{n-2}$

_____

_____

_____

**9** The graph of $ax^2 + bx - 6$ passes through the points $(-3, 9)$ and $(2, 4)$. Find the values of $a$ and $b$, and write the equation for the parabola.

_____

_____

_____

_____

_____

_____

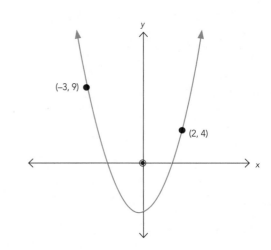

**10** Show that $\dfrac{2}{3(15x^2 + x - 2)} \div \dfrac{1}{9x^2 - 1}$ can be written as $\dfrac{ax + b}{cx + d}$, where a, b, c, d are integers.

_____

_____

_____

_____

PHOTOCOPYING OF THIS PAGE IS RESTRICTED UNDER LAW.    ISBN: 9780170477680

 Answers

## Number (pp. 6–11)
### Prime factorisation (p. 8)

**1**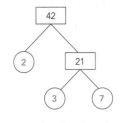
$42 = 2 \times 3 \times 7$

**2**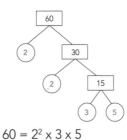
$60 = 2^2 \times 3 \times 5$

**3**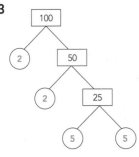
$100 = 2^2 \times 5^2$

**4**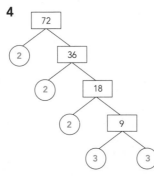
$72 = 2^3 \times 3^2$

### Ratio (pp. 9–11)

**1** 2:3      **2** 5:8

**3** 13:20      **4** 1:14

**5** 0.7:1      **6** 2.25:1

**7** 1      **8** 35

**9** 15      **10** 11

**11** 9 cm      **12** 6000:30000 = 1:5

**13** 75:60:45      **14** 6:10:15

**15** 96:16 = 6:1

**16**

2:11:3

**17** Rectangle 8 x12
Square 10 x 10
Area 96:100 = 24:25

**18** Angle ADB = 22°
Angle CDB = 33°
Angle DBC = 180° – 74 – 33
= 73°

**19** $2[(h \times 4h) + (h \times 3h) + (3h \times 4h)] = 237.5$
$38h^2 = 237.5$
$h = 2.5$
$d = 7.5$
$l = 10$

**20** $\pi \times 5^2 \times \dfrac{45}{360} \times 6 = \dfrac{\pi \times r^2 - \pi \times 5^2}{4}$
$75\pi = \pi r^2 - 25\pi$
$100\pi = \pi r^2$
$r^2 = 100$
$r = \pm 10$
Radius can't be negative, so $r = 10$ cm.

## Algebra (pp. 12–99)
### Powers and roots (pp. 15–22)
#### Multiplying powers (p. 15)

**1** $d^6$      **2** $p^{18n}$

**3** $6y^3$      **4** $e^8$

**5** $15t^{8 + 2n}$      **6** $q^{2.5}$

**7** $-2w^7$      **8** $z^{3.1}$

**9** $35a^{2d + 3}b^{6 + d}$      **10** $12d^5$

**11** $48k^{5.5}$      **12** $-33v^{14}$

**13** $p^{4.1}$      **14** $a^{6n}$

#### Dividing powers (p. 16)

**1** $a^{10}$      **2** $z^8$

**3** $\dfrac{1}{d^{4m}}$ or $d^{-4m}$      **4** $6z^{7.5}$

**5** $\dfrac{1}{2z^6}$ or $\dfrac{1}{2}z^6$      **6** $-5m^5$

**7** $p^{7 - 2x}q^{3x - 1}$      **8** $abc^6$

**9** $\dfrac{a^2}{b}$ or $a^2b^{-1}$      **10** $\dfrac{9x^2}{y^3}$ or $9x^2y^{-3}$

**11** $\dfrac{1}{5j^3k}$ or $\dfrac{1}{5}j^{-3}k^{-1}$      **12** $\dfrac{-y^5z^2}{3}$

**13** $\dfrac{y^5}{2z^4}$ or $\dfrac{1}{2}y^5z^{-4}$      **14** $\dfrac{2a^5b}{7c^5}$ or $\dfrac{2}{7}a^5bc^{-6}$

#### Powers of powers (p. 17)

**1** $y^{30}$      **2** $16z^{20}$

**3** $p^8$      **4** $-q^{10}$

**5** $4x^{6n}$      **6** $-27a^{15}$

**7** $-128b^{15}c^{21}$      **8** $125p^{3b}q^{9b}r^{6b}$

**9** $64a^{12}b^3c^6$      **10** $8d^{24}$

**11** $\dfrac{1}{81}x^{20}$ or $\dfrac{x^{20}}{81}$      **12** $-\dfrac{64}{125}g^6$ or $-\dfrac{64g^6}{125}$

#### Putting it together so far (p. 18)

**1** 1      **2** $\dfrac{16}{a^2}$

**3** $-\dfrac{x^6}{27}$ or $-\dfrac{1}{27}x^6$      **4** $-\dfrac{1}{25b^6}$

**5** $192b^7c^{17}$      **6** $\dfrac{27x^{11}}{yz}$

ISBN: 9780170477680    PHOTOCOPYING OF THIS PAGE IS RESTRICTED UNDER LAW.

**7** $\frac{1}{4}x^6$ or $\frac{x^6}{4}$ or $0.25x^6$    **8** $\frac{-x^6}{y^9}$

**9** $p^4q^3$    **10** $\frac{5y^7}{z^2}$

### Expressions with a common base (p. 19)

**1** $3^{2n-3}$    **2** $2^{4+3n}$

**3** $1$    **4** $3^{2n}$

**5** $5 \times 2^{5n}$    **6** $3^{n^2-3n}$ or $3^{n(n-3)}$

**7** $2^{2n^2-6n}$ or $2^{2n(n-3)}$    **8** $3^{3n+5}$

### Fractional powers and roots (p. 20)

**1** $a^{\frac{1}{4}}$    **2** $x^{\frac{2}{3}}$

**3** $9x^{\frac{1}{2}}$    **4** $5y^{\frac{1}{3}}$

**5** $b^{\frac{7}{2}}$    **6** $11a^{\frac{2}{3}}$

**7** $\sqrt[3]{x}$    **8** $\sqrt[5]{y^2}$

**9** $7\sqrt{a}$    **10** $\sqrt{7a}$

**11** $3^7\sqrt{x^4}$    **12** $12\sqrt{b^9}$

### Negative powers (p. 21)

**1** $\frac{1}{4}$ or $0.25$    **2** $\frac{8}{5}$ or $1\frac{3}{5}$ or $1.6$

**3** $\frac{1}{a^3}$    **4** $\frac{x}{y^2}$

**5** $\frac{4}{x^7}$    **6** $\frac{y^2}{x^4z^3}$

**7** $\frac{1}{25b^2}$    **8** $\frac{1}{x^5y^5}$

**9** $\frac{1}{a^6b^3}$    **10** $\frac{2}{x^3y^2}$

**11** $\frac{1}{16b^2c^6}$    **12** $\frac{4}{9x^2}$

**13** $\frac{1}{a^3}$    **14** $\frac{2y^4}{x^4}$

### Equations with powers (p. 22)

**1** $n = 1$    **2** $n = 8$

**3** $n = 4$    **4** $n = -\frac{1}{2}$

**5** $n = 1$    **6** $n = \frac{1}{2}$

**7** $p = 9q - 2$    **8** $p = 2q - 5$

## Algebraic fractions (pp. 23–27)

### Multiplying and dividing algebraic fractions (pp. 23–24)

**1** $\frac{b^2}{10}$    **2** $\frac{7}{4}$

**3** $\frac{3x^2}{y}$    **4** $\frac{3b(b+1)}{14}$

**5** $4x$    **6** $\frac{2ab}{15}$

**7** $\frac{1}{2b^2}$    **8** $\frac{3y^3}{z}$

**9** $\frac{8(2-b)}{5(b+4)}$    **10** $\frac{10(x-6)}{21(x-4)}$

**11** $\frac{3y}{2z}$    **12** $\frac{18y}{z^2}$

### Adding and subtracting algebraic fractions (pp. 25–26)

**1** $\frac{3y}{4}$    **2** $\frac{a+b}{5}$

**3** $\frac{3x}{4}$    **4** $\frac{2x^2-7}{3}$

**5** $\frac{5+2y}{10}$    **6** $\frac{10x}{11}$

**7** $\frac{17x}{6}$    **8** $\frac{8c+9}{10}$

**9** $\frac{19x+8}{20}$    **10** $\frac{8x-6}{21}$

**11** $\frac{7x}{2}$    **12** $\frac{31x}{20}$

**13** $-\frac{13a}{18}$    **14** $\frac{15ab-8a}{24}$

**15** $\frac{12-x}{6}$    **16** $\frac{13x+4}{20}$

**17** $\frac{7y-2}{4}$    **18** $\frac{7y-4x-3}{7}$

### Simplifying algebraic fractions (p. 27)

**1** $\frac{15x+1}{y}$    **2** $\frac{4-p}{3}$

**3** $\frac{a^2-1}{a}$    **4** $\frac{2b}{a-3b}$

**5** $\frac{x^2y}{3x-y}$    **6** $\frac{2b^2}{1-7b}$

**7** $\frac{q^2+3p}{2}$    **8** $\frac{4q-1}{2pq}$

**9** $\frac{2x^2y-3y}{4x+5xy^2}$ or $\frac{y(2x^2-3)}{x(4+5y^2)}$

**10** $\frac{7xy^2+2y-3}{y^2}$

**11** $\frac{3a^2-ab+2}{b}$

**12** $\frac{y(2x^2z-1)}{3z+y-2xy}$

## Substitution (p. 28)

**1** $-72$    **2** $60$

**3** $1$    **4** $28$

**5** $6$    **6** $6$

**7** $11.25\pi$    **8** $2.25 - 1.125\pi$

**9** $-3.5$    **10** $-0.125$

## Solving equations (pp. 29–36)

**1** $x = -27$    **2** $x = 2$

**3** $x = 34$    **4** $x = 3$

**5** $x = 2.25$    **6** $x = 17$

**7** $x = 24$    **8** $x = 1995$

**9** $x = -4.5$    **10** $x = -22$

**11** $x = \frac{2}{3}$    **12** $x = 21$

**13** $x = 2$    **14** $x = 5$ or $x = -5$

### Forming and solving linear equations (pp. 31–36)

**1** $2((4x - 3) + (x + 1)) = 24$
$x = 2.8$ cm
Sides: 3.8 cm and 8.2 cm

**2** $(x + 12) + (2x - 15) + (4x + 8) = 180$
$x = 25$
Angles 37°, 35° and 108°

**3** $(5x - 82) + x + (4x + 2) = 360$
$x = 44°$
$4x + 2 = 178°$ so bc is not a diameter.

PHOTOCOPYING OF THIS PAGE IS RESTRICTED UNDER LAW.    ISBN: 9780170477680

**4** $6x - 23 = 4x + 2$
$x = 12.5$ so each side = 52 cm
Perimeter is 156 cm so it is bigger than 150 cm.

**5** $1.8x + 1.8x + x + x = 784$
$x = 140$ cm
Width $x = 252$ cm

**6** $2(8x + 2) + 2(5x) = 58 + 58 + 44$
$16x + 4 + 10x = 160$
$26x + 4 = 160$
$26x = 156$
$x = 6$
Rectangle is 30 by 50.

**7** $6x - 10 = 5x + 3$
$x = 13°$
$\therefore 5x + 3 = 68$
$6y - 2x = 180 - 68$
$6y - 26 = 112$
$6y = 138$
$y = 23°$

**8** $y = x + 7$  **9** $y = 2x + 5$
**10** $y = 3x - 4$  **11** $y = -2x + 3$
**12** $y = -0.5x + 11$  **13** $y = 1.5x - 17$
**14** $y = 1.5x - 17$  **15** $y = -0.8x - 23$

**16** $m = \dfrac{2p - (-3p)}{5}$
$= p$
$y^2 - (-3p) = p(x - 4)$
$y + 3p = px - 4p$
$y = px - 7p$

**17** $m = \dfrac{17 - 8}{-p - 2p}$
$= \dfrac{-3}{p}$
$y^2 - 8 = \left(\dfrac{-3}{p}\right)(x - 2p)$
$y = -\dfrac{-3}{p}x + 14$

**18** $y = -\dfrac{1}{3}x$  **19** $y = \dfrac{2}{3}x - 12$

**20** $y = -\dfrac{1}{4}x + 5$  **21** $y = \dfrac{5}{6}x - 7$

**22** $m_1 = \dfrac{7}{2}$
Perpendicular $\Rightarrow m_2 = \dfrac{7}{2}$
$y = \dfrac{7}{2}x - 2$

**23** $y = \dfrac{3}{8}x - \dfrac{15}{8} \Rightarrow m_1 = \dfrac{3}{8}$
Perpendicular $\Rightarrow m_2 = -\dfrac{8}{3}$
$\therefore y - 6 = -\dfrac{8}{3}(x - 21)$
$y = -\dfrac{8}{3}x + 62$

**24** $y = -\dfrac{1}{3}x + \dfrac{7}{3} \Rightarrow m_1 = -\dfrac{1}{3}$
Perpendicular $\Rightarrow m_2 = a = 3$

**25** $by = -5x - 3 \Rightarrow m_1 = -\dfrac{5}{b}$
$y = -2x - 5 \Rightarrow m_2 = -2$
Parallel $\Rightarrow \dfrac{-5}{b} = -2$
$\therefore b = \dfrac{5}{2}$

**26** From A: $y = \dfrac{2}{3}x - \dfrac{7}{12} \Rightarrow m_1 = \dfrac{2}{3}$
Parallel $\Rightarrow m_1 = \dfrac{2}{3}$
From A: $y = \dfrac{p}{-9}x - \dfrac{10}{9}$
$\therefore \dfrac{p}{-9} = \dfrac{2}{3} \Rightarrow p = -6$

**27** From A: $y = -\dfrac{2}{5}x + 1 \Rightarrow m_1 = -\dfrac{2}{5}$
Perpendicular $\Rightarrow m_2 = \dfrac{5}{2}$
From B: $py = 15x - 18$
$\therefore \dfrac{5}{2} = \dfrac{15}{p} \Rightarrow p = 6$

## Inequalities (pp. 37–42)
### Understanding inequation signs (p. 37)
**1** $6 < 11$  **2** $4 > -2$
**3** $-9 < -8$  **4** $-0.5 < 0$
**5** $x$ is less than 9.  **6** $x$ is less than $-2$.
**7** $x$ is greater than 1 and less than 14
**8** $x$ is greater than or equal to $-7$.
**9** $x$ is less than or equal to 0.
**10** $x$ is greater than or equal to 5 and less than or equal to 198.
**11** $x < 11$  **12** $x \geq 2$
**13** $-1 \leq x \leq 4$  **14** $x \leq 7$
**15** $5 \leq x < 8$

### Using inequality expressions (pp. 38–39)
**1** $A = 1, B = 3$ or $A = 3, B = 5$
**2** $y \leq 24$
**3** $a > c$ and $c < b$
**4**

|  | Always true | Sometimes true | Never true |
|---|---|---|---|
| $4 < x$ | ✗ | ✗ | ✓ |
| $x + 1 < 0$ | ✗ | ✓ | ✗ |
| $x^2 > 10$ | ✗ | ✗ | ✓ |

**5** **a** 5.5 cm $\leq L <$ 6.5 cm
1.5 cm $\leq H <$ 2.5 cm
**b** 8.25 cm² $\leq A <$ 16.25 cm²
**c** 14 cm $\leq P <$ 18 cm

**6** **a** 21  **b** 8
**c** 21.9 so it would increase.

**7 a** $11.5\text{ m} \le d < 12.5\text{ m}$
  **b** $5.75\text{ m} \le r < 6.25\text{ m}$
  **c** $33.0625\pi\text{ m}^2 \le A < 42.25\pi\text{ m}^2$
    Area for $r = 6$ m is $36\pi$ m², which is $6.25\pi$ m²
    above the minimum area and $2.937\pi$ m² below
    the maximum area.

**8** $\pi \times 5^2 - \pi \times 4.8^2 \le A \le \pi \times 9^2 - \pi \times 2.5^2$
  or $1.96\pi\text{ cm}^2 \le A \le 74.75\pi\text{ cm}^2$

### Solving linear inequations (pp. 40–41)

**1** $x < 4$. So $x$ is less than 4.
**2** $x \le 13$. So $x$ is less than or equal to 13.
**3** $x > 8$. So $x$ is greater than 8.
**4** $x \ge 2$. So $x$ is greater than or equal to 2.
**5** $x > -21$. So $x$ is greater than –21.
**6** $x \le 13$. So $x$ is less than or equal to 13.
**7** $x \ge 11$. So $x$ is greater than or equal to 11.
**8** $x > 24$. So $x$ is greater than 24.
**9** $x \le 10$. So $x$ is less than or equal to 10.
**10** $x > -16$. So $x$ is greater than –16.
**11** $4 < x < 12$. So $x$ is greater than 4 and less than 12.
**12** $x > \dfrac{17}{2}$. So $x$ is greater than $\dfrac{17}{2}$.
**13** $x = 2$      **14** $y = 4$

### Forming and solving linear equations (p. 42)

**1** $1 + 3x < 5x - 8$
  $x > 4.6$ and a multiple of 3.
  $x = 6$
**2** $3(2x + 7) > 11x - 9$
  $x < 6$
**3** $2(5x + 1) \ge 4(3x - 1)$
  $x \le 3$
  Maximum is 3.
**4** $\dfrac{1}{2} \times 7 \times 2(5x - 1) < \dfrac{1}{2} \times 11(6x + 5)$

       $14(5x - 1) < 11(6x + 5)$
               $x < 17.25$
               Maximum is 17.
**5** $\dfrac{3a}{2} \times 5 \le 29$

       $a \le 3.8\dot{6}$ and an interger
       $a \le 3$ cm
       Maximum is 3.

## Simultaneous equations (pp. 43–48)

### Substitution (pp. 43–44)

**1** $(5, 6)$      **2** $(2, 8)$
**3** $(11, 8)$      **4** $(3, 15)$
**5** $(4, 18)$      **6** $(2, 6)$
**7** $(5, -2)$      **8** $(-88, -22)$
**9** $(5, -1)$      **10** $(-4, 11)$

### Elimination (pp. 45–46)

**1** $(2, 4)$      **2** $(1, 4)$
**3** $(4, -4)$      **4** $(0, 2)$
**5** $(-6, 1)$      **6** $(2, 2)$
**7** $(-5, 11)$      **8** $(7, 1.5)$
**9** $(-1, 2)$      **10** $(-4, 6)$

### Forming and solving simultaneous equations (pp. 47–48)

**1** $x + y = 23$
  $x - y = 9$
     $x = 16, y = 7$.
  Larger number = 16.
**2** $2b + 2h = 150$
       $b = h + 15$
       $h = 30$ cm $b = 45$ cm
**3** $x + y = 90$
  $x - y = 26$
     $x = 58°, y = 32°$
**4** $y = -x + 5$
  $y = 0.5x + 2$
  $x = 2, y = 3$
  They intersect at $(2, 3)$.
**5** $3(x + y) = 120$
  $\dfrac{1}{2}(x - y) = 30$
       $x = 50, y = -10$
**6** $5x - 6y = 90$
  $2x + 3y = 90$
       $x = 30, y = 10$
**7** $6a + 8b = 34$
  $10a + 4b = 38$
       $a = 3$
       $b = 2$
  Rectangle X is 8cm by 9 cm.
  Rectangle Y is 15 cm by 4 cm.

## Challenge 1 (p. 49)

**1** $2x - 6 - 4x + 2 < 14 + 7x$
       $-2x - 4 < 14 + 7x$
         $-9x < 18$
           $x > -2$
**2** $5(5x - 2) + 4(2x + 3) = 200$
     $25x - 10 + 8x + 12 = 200$
               $33x = 198$
                  $x = 6$
**3** $5x - (-2)2 = \sqrt{\sin 90°}$
       $5x - 4 = 1$
       $5x = 1 + 4$
         $x = \dfrac{1 + 4}{5}$
         $x = 1$
**4** $2^{x+4} < 2^{3x}$
     $x + 4 < 3x$
         $4 < 2x$
         $x > 2$
**5** $(5, 2)$
  $x - y = 3$

  $7x - 3y = 29$    $(\times 7)$
  $4x - 7y = 6$     $(\times 3)$

  $49x - 21y = 203$
  $12x - 21y = 18$

PHOTOCOPYING OF THIS PAGE IS RESTRICTED UNDER LAW.    ISBN: 9780170477680

$37x = 185$

$x = 5$

$y = 2$

$x - y = 3$

**6** $5^{2x} \times 5^y = 5^3$

$2x + y = 3$

$2^{3x} \times 2^{2y} = 27$

$3x + 2y = 7$

$2x + y = 3$ ①

$3x + 2y = 7$ ②

$4x + 2y = 6$ ① × 2

$3x + 2y = 7$

$x = -1$

$y = 5$

## Quadratic expressions (pp. 50–86)

### Expanding quadratic expressions (pp. 50–51)

**1** $x^2 + 11x + 18$

**2** $x^2 + x - 20$

**3** $x^2 - 4x - 21$

**4** $x^2 - 12x + 32$

**5** $2x^2 - 5x - 12$

**6** $x^2 + 12x + 27$

**7** $x^2 + 18x + 81$

**8** $x^2 - 14x + 49$

**9** $3x^2 - 22x - 16$

**10** $12x^2 - 14x - 10$

**11** $4x^2 - 28x + 49$

**12** $x^2 - 64$

**13** $18x^2 - 33x + 12$

**14** $x^2 - 8x + 16$

**15** $-24x^2 - 13x + 2$

**16** $9x^2 - 12x + 3$

**17** $12x^2 - 26x + 10$

**18** $-9x^2 + 81$

### Factorising quadratic expressions (pp. 52–56)

**1** $(x + 1)(x + 7)$

**2** $(x + 1)(x + 6)$

**3** $(x + 3)(x + 9)$

**4** $(x + 8)(x - 2)$

**5** $(x + 14)(x + 5)$

**6** $(x - 4)(x - 5)$

**7** $(x + 2)(x - 11)$

**8** $(x - 3)^2$

**9** $(x + 6)^2$

**10** $(x + 6)(x - 7)$

**11** $(x - 4)^2$

**12** $(x + 9)(x - 1)$

**13** $(x + 9)(x - 9)$

**14** $(5 + x)(5 - x)$

**15** $(3 + 2x)(3 - 2x)$

**16** $(1 + 5x)(1 - 5x)$

**17** $(9 + x)(3 + x)$

**18** $(3 - x)(7 - x)$

**19** $2(x + 1)(x + 7)$

**20** $5(x + 1)(x + 3)$

**21** $3(x - 1)(x + 4)$

**22** $4(x - 5)(x - 8)$

**23** $6(x + 2)^2$

**24** $10(x - 3)(x + 3)$

**25** $-2(x - 7)^2$

**26** $3(x - 1)(x + 1)$

**27** $4(x + 2)(x - 3)$

**28** $5(4 - x)(4 + x)$

**29** $(2x + 1)(x + 5)$

**30** $(2x + 3)(x + 2)$

**31** $(2x - 5)(x - 1)$

**32** $(2x - 5)(x + 2)$

**33** $(3x - 1)(x - 4)$

**34** $(4x - 1)(2x + 3)$

**35** $(4x + 3)(x - 9)$

**36** $(3x + 5)(2x - 3)$

**37** $(3x - 2)(4x + 3)$

**38** $(3x - 1)(3x - 5)$

**39** $(2x - 1)^2$

**40** $2(3x - 1)(3x + 1)$

### Simplifying quadratic fractions (pp. 57–58)

**1** $\dfrac{2}{x + 5}$

**2** $\dfrac{x - 3}{3}$

**3** $x$

**4** $4(x + 2)$

**5** $x + 5$

**6** $x + 3$

**7** $\dfrac{1}{x - 1}$

**8** $\dfrac{5}{x - 4}$

**9** $\dfrac{x + 9}{x - 8}$

**10** $\dfrac{x - 1}{x - 7}$

**11** $x - 1$

**12** $\dfrac{-1}{3 + x}$

**13** $\dfrac{-(7 + x)}{3x + 5}$

**14** $\dfrac{9x + 4}{5 - x}$

**15** $\dfrac{x - 3}{5x + 4}$

**16** $\dfrac{7x - 2}{4x + 1}$

**17** $\dfrac{x + 1}{3x}$

**18** $\dfrac{-x}{2x + 5}$

**19** $\dfrac{x + 6}{5 - x}$ or $\dfrac{x + 6}{-(x - 5)}$

**20** $\dfrac{1}{2(x + 5)}$ or $\dfrac{1}{2x + 10}$

### Solving quadratic equations (pp. 59–63)

**1** $x = -8$ or $2$

**2** $x = -4$ or $6$

**3** $x = -7$ or $-2$

**4** $x = 4$ or $1$

**5** $x = -5$ or $9$

**6** $x = 6$

**7** $x = -8$

**8** $x = \dfrac{1}{2}$

**9** $x = -3$ or $2$

**10** $x = -6$ or $3$

**11** $x = 0$ or $2$

**12** $x = 0$ or $-4$

**13** $x = 0$ or $\dfrac{3}{4}$

**14** $x = 0$ or $-\dfrac{7}{2}$

**15** $x = 0$ or $4$

**16** $x = 0$ or $\dfrac{3}{4}$

**17** $x = -8$ or $-1$

**18** $x = 2$ or $-6$

**19** $x = 5$ or $1$

**20** $x = 8$ or $0$

**21** $x = 2$ or $-8$

**22** $x = -5$

**23** $x = -6$ or $0$

**24** $x = -9$ or $4$

**25** $x = 8$

**26** $x = -6$ or $7$

**27** $x = -9$

**28** $x = -3$ or $3$

**29** $x = -6$ or $12$

**30** $x = -8$ or $10$

**31** $x = -4$ or $-\dfrac{1}{3}$

**32** $x = \dfrac{1}{2}$ or $\dfrac{5}{2}$

**33** $x = 4$ or $\dfrac{7}{2}$

**34** $x = 2$ or $\dfrac{1}{4}$

**35** $x = 8$ or $-3$

**36** $x = 10$ or $-7$

**37** $x = -9$ or $3$

**38** $x = 6$ or $3$

**39** $x = 11$ or $-4$

**40** $x = -8$ or $3$

**41** $x = 14$ or $-3$

**42** $x = -6$ or $-5$

### Forming quadratic equations (pp. 64–71)

**1** Difference $= (2x - 1)^2 - (x + 3)^2$

$= (4x^2 - 4x + 1) - (x^2 + 6x + 9)$

$= 3x^2 - 10x - 8$

**2** Area $= (4z - 1)^2 - \dfrac{1}{2} \times 2z(z + 1)$

$= 16z^2 - 8z + 1 - z^2 - z$

$= 15z^2 - 9z + 1$

**3** $2x(x + 9) = 2(x + 1)(x + 5)$

$2x^2 + 18x = 2x^2 + 12x + 10$

$3x - 5 = 0$

**4**  Area = 22y × 16y – 14 × 6y
     = $352y^2 - 84y^2$
     = $268y^2$

**5**  $\dfrac{(3x + 2)(2x - 1)}{2}$ × (x + 5) = 38

     $(5x + 1)(x + 5) = 76$
     $5x^2 + 26x - 71 = 0$

**6**  Area = π × $(2x - 1)^2$ – π × $(z - 1)^2$
     = $\pi((4z^2 + 4z + 1) - (z^2 - 2z + 1))$
     = $\pi(3z^2 + 6z)$
     = $3\pi z(z + 2)$

**7**  A = $\dfrac{\pi}{2}(x + 1)^2 - \pi\left(\dfrac{x + 1}{2}\right)^2$

     = $\dfrac{\pi}{2}(x + 1)^2 - \dfrac{\pi}{4}(x + 1)^2$

     = $\dfrac{\pi}{4}(x + 1)^2$ or $\dfrac{\pi}{4}(x^2 + 2x + 1)$

Half the semicircle is shaded.

**8**  SA = $2\pi r^2 + 2\pi rh$
     = $2\pi(a + 3)^2 + 2\pi(a + 3) \times 2(a + 3)$
     = $2\pi(3a^2 + 18a + 27)$
     = $6\pi(a^2 + 6a + 9)$ or $6\pi(a + 3)^2$

**9**

| x | 1 | 2 | 3 | 4 | 5 |
|---|---|---|---|---|---|
| y | 4 | 7 | 12 | 19 | 28 |
| First difference | | 3 | 5 | 7 | 9 |
| Second difference | | | 2 | 2 | 2 |
| $x^2$ | 1 | 4 | 9 | 16 | 25 |
| $y - x^2$ | 3 | 3 | 3 | 3 | 3 |

Equation: $y = x^2 + 3$

**10**

| x | 1 | 2 | 3 | 4 | 5 |
|---|---|---|---|---|---|
| y | 3 | 8 | 15 | 24 | 35 |
| First difference | | 5 | 7 | 9 | 11 |
| Second difference | | | 2 | 2 | 2 |
| $x^2$ | 1 | 4 | 9 | 16 | 25 |
| $y - x^2$ | 2 | 4 | 6 | 8 | 10 |

Equation: $y = x^2 + 2x$

**11**

| x | 1 | 2 | 3 | 4 | 5 |
|---|---|---|---|---|---|
| y | 2 | 8 | 16 | 26 | 38 |
| First difference | | 6 | 8 | 10 | 12 |
| Second difference | | | 2 | 2 | 2 |
| $x^2$ | 1 | 4 | 9 | 16 | 25 |
| $y - x^2$ | 1 | 4 | 7 | 10 | 13 |

Equation: $y = x^2 - 3x - 2$

**12**

| x | 1 | 2 | 3 | 4 | 5 |
|---|---|---|---|---|---|
| y | 7 | 24 | 51 | 88 | 135 |
| First difference | | 17 | 27 | 37 | 47 |
| Second difference | | | 10 | 10 | 10 |
| $5x^2$ | 5 | 20 | 45 | 80 | 125 |
| $y - 5x^2$ | 2 | 4 | 6 | 8 | 10 |

Equation: $y = 5x^2 + 2x$

**13**

| x | 1 | 2 | 3 | 4 | 5 |
|---|---|---|---|---|---|
| y | –2 | –1 | 4 | 13 | 26 |
| First difference | | 1 | 5 | 9 | 13 |
| Second difference | | | 4 | 4 | 4 |
| $2x^2$ | 2 | 8 | 18 | 32 | 50 |
| $y - 2x^2$ | –4 | –9 | –14 | –19 | –24 |

Equation: $y = 2x^2 - 5x + 1$

**14**

| x | 1 | 2 | 3 | 4 | 5 |
|---|---|---|---|---|---|
| y | 2 | 6 | 8 | 8 | 6 |
| First difference | | 4 | 2 | 0 | –2 |
| Second difference | | | –2 | –2 | –2 |
| $-1x^2$ | –1 | –4 | –9 | –16 | –25 |
| $y + 2x^2$ | 3 | 10 | 17 | 24 | 31 |

Equation: $y = -x^2 + 7x - 4$

**15 a**  Product = $(2x + 1)(2x + 3)$
           = $4x^2 + 8x + 3$

 **b**  For 11 and 13, x = 5.
     Product = $4x^2 + 8x + 3$
          = 100 + 40 + 3
          = 143

**16 a**  Sum = $(2x + 1) + (2x + 1)^2$
         = $2x + 1 + 4x^2 + 4x + 1$
         = $4x^2 + 6x + 2$

 **b**  For 15, x = 7
     Sum = $4x^2 + 6x + 2 = 4 \times 7^2 + 4(7) + 2$
         = 240
     Sum = $15 + 15^2$
         = 240

**17**  Product = $(2x + 1)(2x + 5)$
              = $4x^2 + 12x + 5$

**18**  Square of the sum = $[(2x + 1) + (2x + 3)]^2$
                   = $[4x + 4]^2$
                   = $16x^2 + 32x + 16$

**19**  Sum = $(2x)^2 + (2x + 2)^2 = 4x^2 + 4x^2 + 8x + 4$
              = $8x^2 + 8x + 4$

PHOTOCOPYING OF THIS PAGE IS RESTRICTED UNDER LAW.     ISBN: 9780170477680

**20**   Difference = $((n + 1)^2 + 3) - (n^2 + 3)$
               = $n^2 + 2n + 1 + 3 - n^2 - 3$
               = $2n + 1$

**21**   Sum = $((n + 1)^2 - 3(n + 1) + 2) + (n^2 - 3n + 2)$
           = $n^2 + 2n + 1 - 3n - 3 + 2 + n^2 - 3n + 2$
           = $2n^2 - 4n + 2$

**Forming and solving quadratic equations (pp. 72–74)**

**1**   $(x + 6)(x - 2) - 84$
    $x^2 + 4x - 12 = 84$
    $x^2 + 4x - 96 = 0$
  $(x + 12)(x - 8) = 0$
            $x = -12$ or $8$
x cannot be negative, so x must equal 8.
Base = 6 m
Height = 14 m

**2**   $\pi r^2 = 50$
   $r^2 = \dfrac{50}{\pi}$
   $r = \sqrt{\dfrac{50}{\pi}}$
   $r = \pm 3.99$ (2 dp)
   $r = 3.99$ m (r can only be positive)
Diameter = 7.98 m (2 dp)

**3**        $x^2 - 132 = x$
    $x^2 - x - 132 = 0$
 $(x - 12)(x + 11) = 0$
          $x = 12$ or $-11$
x must be a positive square number, so it cannot be −11, and must be 12.

**4**   $(x - 4)^2 + x^2 = 10$
 $2x^2 - 8x + 16 = 10$
  $2x^2 - 8x + 6 = 0$
   $x^2 - 4x + 3 = 0$
 $(x - 1)(x - 3) = 0$
        $x = 1$ or $3$
Check answers: $x = 1 \Rightarrow (-1)^2 + 3^2 = 10$
                $x = 3 \Rightarrow (3)^2 + 1^2 = 10$
So x = 1 or 3.

**5**      $x(x + 11) = 42$
   $x^2 - x - 132 = 0$
 $(x - 12)(x + 11) = 0$
          $x = 12$ or $-11$
x cannot be negative, so x = 3.
Perimeter = 2(3) + 2(14) = 34 cm

**6**   $\dfrac{1}{2}(2x - 4)(x + 4) = 16$
    $(x - 2)(x + 4) = 16$
    $x^2 - 2x - 24 = 0$
   $(x - 4)(x + 6) = 0$
         $x = -6$ or $4$
x cannot be negative, so x = 4.
∴ Perpendicular sides must be 4 and 8.

**7**   $(4x + 1)(3x + 1) - 2x(x + 1) = 51$
  $12x^2 + 7x + 1 - 2x^2 - 2x = 51$
     $10x^2 + 5x - 50 = 0$
      $2x^2 + x - 10 = 0$
   $(x - 2)(2x + 5) = 0$
               $x = 2$ or $-\dfrac{5}{2}$
x cannot take a negative value, so x = 2 m

**8**     $\pi \times 6^2 - r^2 = 62.83$
       $-\pi r^2 = -50.267$
         $r^2 = 16.00$
          $r = \pm 4$
          $r = 4$ m (r cannot be positive)
Radius of shaded ring = 6 m − 4 m
                     = 2 m

**9**   $x^2 + ax + 6 = 0$
Factors of 6: (1,6) or (2,3).
+ 6 ⇒ solutions have same signs.
Difference = 1 ⇒ factors must be 2 and 3.
∴ $(x + 2)(x + 3) = 0$ or $(x - 2)(x - 3) = 0$
a is positive ⇒ $(x + 2)(x + 3) = 0$
∴ a = 5 and x = −2 or −3.

**10**   $x^2 + ax - 12 = 0$
Factors of 12: (1,12), (2,6) or (3,4).
− 12 ⇒ solutions have different signs.
Difference = 1 ⇒ factors must be 3 and 4.
∴ $(x + 3)(x - 4) = 0$ or $(x - 3)(x + 4) = 0$
a is negative ⇒ $(x + 3)(x - 4) = 0$
∴ a = −1 and x = −3 or 4.

**11**   $x^2 + ax - 15 = 0$
Factors of 15: (1,15) or (3,5).
− 15 ⇒ solutions have different signs.
Difference = 8 ⇒ factors must be 3 and 5.
∴ $(x + 3)(x - 5) = 0$ or $(x - 3)(x + 5) = 0$
a is positive ⇒ $(x - 3)(x + 5) = 0$
∴ a = 2 and x = 3 or −5.

**12**   $x^2 + ax + 36 = 0$
Factors of 36: (1, 36), (2,18), (3, 12), (4,9) and (6,6).
+ 36 ⇒ solutions have same signs.
Solutions are ±p ⇒ factors must be 6 and 6.
∴ $(x + 6)(x + 6) = 0$ or $(x - 6)(x - 6) = 0$
a is negative ⇒ $(x - 6)(x - 6) = 0$
∴ a = −12 and x = 6.
a is positive ⇒ $(x + 6)(x + 6) = 0$
∴ a = 12 and x = −6.

**Plotting quadratic equations (pp. 75–77)**

**1**

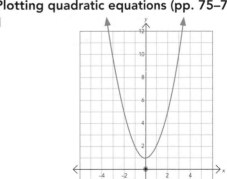

The vertex is minimum/~~maximum~~ at (0, 1)

ISBN: 9780170477680    PHOTOCOPYING OF THIS PAGE IS RESTRICTED UNDER LAW.

**2**

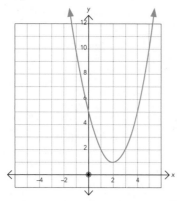

The vertex is minimum/~~maximum~~ at (2, 1)

**3**

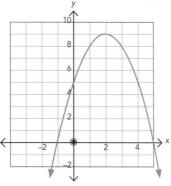

The vertex is ~~minimum~~/maximum at (2, 9)

**4**

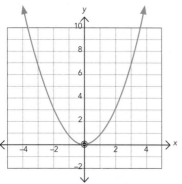

The vertex is minimum/~~maximum~~ at (0, 0)

**5**

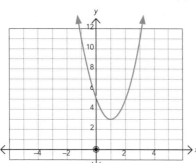

The vertex is minimum/~~maximum~~ at (1, 3)

**6**

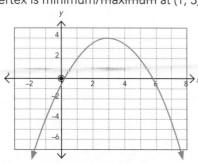

The vertex is ~~minimum~~/maximum at (3, 4)

Writing equations for parabolas (pp. 78–83)

**1**  $y = (x - 3)(x + 2)$     **2**  $y = (x - 2)(x - 6)$
**3**  $y = -(x + 1)(x - 4)$     **4**  $y = x(x + 5)$
**5**  $y = 2(x - 3)(x + 2)$     **6**  $y = -0.5x(x - 6)$
**7**  $y = 0.25(x - 1)(x + 4)$     **8**  $y = -3x(x + 3)$
**9**  $y = (x - 3)^2 + 2$     **10**  $y = (x + 2)^2 + 1$
**11**  $y = -(x - 2)^2 + 5$     **12**  $y = 0.5(x + 4)^2 - 6$
**13**  $y = 2(x - 2)^2 + 3$     **14**  $y = 0.25(x - 4)^2 - 3$
**15**  $y = -\dfrac{1}{3}(x + 4)^2 + 6$     **16**  $y = -4(x - 2)^2 + 5$

Quadratic inequations (p. 84)

**1**  $(x - 2)(x + 5) \le 0$
   $-5 \le x \le 2$
**2**  $(x + 2)(x - 7) > 0$
   $-2 < x < 7$
**3**  $81 < x^2 < 121$
   $9 < x < 11$ and $-11 < x < -9$
**4**  $4 < x^2 < 100$
   $2 < x < 10$ and $-10 < x < -2$

Mixing it up (pp. 85–86)

**1**  Skinnier
**2**  Wider
**3**  **a**  (0, 0) and (5, 0)
   **b**  (0, –5)
**4**  **a**  $x = -3$ and $x = 3.5$
   **b**  $y = -21$
**5**  (6, –1)
   Maximum
**6**  (–3, –1)
   Minimum
**7**  $y = (x - 1)^2 - 5$ or $y = x^2 - 2x - 4$
**8**  $y = -(x + 1)^2 + 3$ or $y = -x^2 - 2x + 2$
**9**  $p = 5$
   $q = 20$
**10**  $c = 16$
**11**  $b = \pm 18$
**12**  **a**  $x^2 + 14x + 49$
   **b**  $x^2 + 2x - 15$ or $x^2 + 14x - 15$
   **c**  $x^2 - 10x - 11$
   **d**  $2x^2 + 5x - 3$ or $2x^2 + x - 3$
**13**  **a**  $y = (x - 3)^2 + 2$
     or $y = x^2 - 6x + 11$
   **b**  $y = (x - 3)^2 + 0$
     or $y = x^2 - 6x + 9$
**14**  **a**  $y = -(x - 4)(x + 1)$
     or $y = (4 - x)(x + 1)$
     or $y = -x^2 + 3x + 4$
   **b**  $y = -(x - 5)(x + 1)$
     or $y = (5 - x)(x + 1)$
**15**  **a**  $y = 2$
   **b**  $y = 3(x + 2)(x + 1)$
**16**  **a**  $y = -12$
   **b**  $y = 0.5(x + 2)(x - 6)$
**17**  $x < -5$ and $x > 3$
**18**  $-2 < x < 7$

PHOTOCOPYING OF THIS PAGE IS RESTRICTED UNDER LAW.  ISBN: 9780170477680

## Challenge 2 (pp. 87–88)

**1**   $5x^2 + 5x - 30 \le 0$
     $5(x - 2)(x + 3) \le 0$
     $-3 \le x \le 2$

**2**   $2x^2 + 10x + 8 = 0$
     $2(x + 1)(x + 4) = 0$
     $x = -1$ or $x = -4$

**3**   $x^2 + 3x - 10 = 0$
     $(x + 5)(x - 2) = 0$
     $x = -5$ or $x = 2$

**4**   $m = 7,\ n = -\dfrac{1}{2}$

     $m - n = 7\dfrac{1}{2}$

**5**   Area $= 3(x + 2) + (2x + 4)(x + 7)$
        $= 3x + 6 + 2x^2 + 4x + 14x + 28$
        $= 2x^2 + 21x + 34$

**6**   $x^2 - 9 > x^2 + x - 12$
     $-9 > x - 12$
      $x < 3$

**7**   $A = 3x^2 - 15x + 6 + 3x$
      $= 3x^2 - 12x + 6$
     $B = 2x^2 - 12x - 14 + x^2 - 1$
      $= 2x^2 - 12x - 15$
     $A = B + 21$

**8**   $(10, 13) \Rightarrow 13 = 100a + 10b + 3$
         $10a + b = 1$    ①
     $(2, -3) \Rightarrow -3 = 4a + 2b + 3$
          $2a + b = -3$    ②
     Subtract ② from ①:
        $8a = 4$
         $a = 0.5$
     Substitute for a in ①: $5 + b = 1$
                    $b = -4$
     Equation: $0.5x^2 - 4x + 3$

**9**   $\dfrac{9x^2 - 4}{3(3x - 2)} \times \dfrac{2x}{x^2 - 3x} = \dfrac{(3x + 2)(3x - 2)}{3(3x - 2)} \times \dfrac{2x}{x(x - 3)}$

                $= \dfrac{3x^2 - 2}{3} \times \dfrac{2}{x - 3}$

                $= \dfrac{6x - 4}{3x - 9}$

     $a = 6,\ b = -4,\ c = 3$ and $d = -9$

## Exponential equations (pp. 89–93)

### Basic exponential equations (p. 89)

**1**   $x = 4$          **2**   $x = 5$
**3**   $x = 3$          **4**   $x = 7$
**5**   $x = 11$        **6**   $x = 0$
**7**   $x = 5$          **8**   $x = 3$

### Plotting exponential graphs (pp. 90–91)

**1**

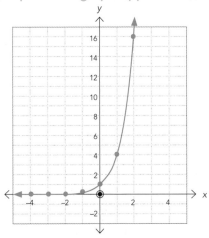

### Translated exponential graphs (pp. 92–94)

**1**   $y = 2^{x-1}$         **2**   $y = 2^x + 2$
**3**   $y = 2^x - 1$        **4**   $y = 2^{x+2}$
**5**   $y = 2^x + 3$        **6**   $y = 2^{x-2}$

**7**

| x | y |
|---|---|
| 0 | –1 |
| 1 | 1 |
| 2 | 7 |

The y intercept is at **(0, –1)**
The equation of the asymptote is
$y = -2$

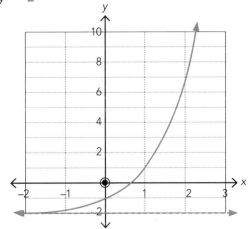

**8**

| x | y |
|---|---|
| –1 | 1 |
| 0 | 4 |
| 1 | 16 |

The y intercept is at **(0, 4)**
The equation of the asymptote is
$y = 0$

ISBN: 9780170477680    PHOTOCOPYING OF THIS PAGE IS RESTRICTED UNDER LAW.     

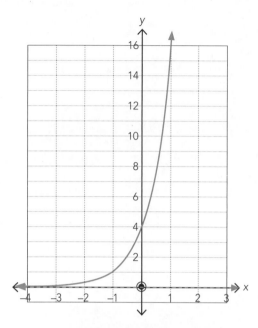

**9**

| x | y |
|---|---|
| 0 | 1 |
| 2 | 2 |
| 4 | 4 |
| 6 | 6 |

The y intercept is at **(0, 1)**
The equation of the asymptote is
**y = 0**

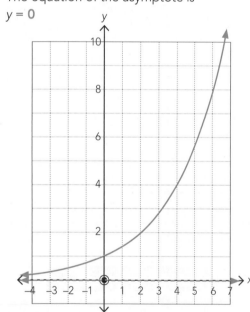

### Challenge 3 (p. 95)

**1**     $x^2 \leq 6x$
    $x(x - 6) \leq 0$
        $0 \leq x \leq 6$

**2**   $2^{x^2} \times 2^{2x} \div 2^3 = 0$
    $x^2 - 2x - 3 = 0$
    $(x - 1)(x + 3) = 0$
            $x = 1$ or $-3$

---

**3**        $2^{x^2} > 2^{3x} \times 2^{10}$
    $x^2 - 3x - 10 > 0$
    $(x + 2)(x - 5) > 0$
            $x < -2$ or $x > 5$

**4**        $3^{x^2} \leq 27 \div 9^x$
        $3^{x^2} \leq 27 \div 3^{2x}$
    $3^{x^2} + 3^{2x} - 3^3 \leq 0$
        $x^2 + 2x - 3 \leq 0$
        $(x + 3)(x - 1) \leq 0$
                $-3 \leq x \leq 1$

**5**   $2^{n + 3} < 57.1$
        $2^5 = 32$ and $2^6 = 64$
    so $n + 3 \leq 5$
        $n \leq 2$

**6**   $3^{2n - 5} > 45.45$
        $3^3 = 27$ and $3^4 = 81$
    so $2n - 5 \geq 4$
        $2n \geq 9$
        $n \geq 5$

**7**        $5^{3n - 2} = 5^{-2n^2}$
    $2n^2 + 3n - 2 = 0$
    $(2n - 1)(n + 2) = 0$
            $n = \dfrac{1}{2}$ or $-2$

**8**     $4^4 \times 4^{p^2} = 4^{4p}$
        $p^2 + 4 = 4p$
    $p^2 - 4p + 4 = 0$
        $(p - 2)^2 = 0$
                $p = 2$

### Rearrangement of expressions (pp. 96–99)

**1   Where the subject appears once (pp. 96–97)**

**1**   $c = 2b - a^2$                **2**   $R = \dfrac{100I}{PT}$

**3**   $y = \dfrac{2x^2}{3}$         **4**   $b = \pm\sqrt{\dfrac{3V}{h}}$

**5**   $m = \dfrac{y - c}{x}$        **6**   $a = \dfrac{v^2 - u^2}{2s}$

**7**   $C = \dfrac{5(F - 32)}{9}$    **8**   $r = \sqrt[3]{\dfrac{3V}{4\pi}}$

**9**   $\theta = \dfrac{360A}{\pi r^2}$   **10**   $u = \dfrac{2s - at^2}{2t}$

**11**   $a = \pm\sqrt{b}$            **12**   $a = b^2$

**2   Where the subject appears twice (pp. 98–99)**

**1**   $b = \dfrac{12}{8 + a}$       **2**   $y = \dfrac{2x - 9}{7}$

**3**   $a = \dfrac{7}{b + 3}$        **4**   $b = \dfrac{5a}{4 + 2a}$

**5**   $y = \dfrac{5x - 3}{2 - x}$   **6**   $b = \dfrac{5 + a}{a - 2}$

**7**   $ab + ac = bc$
    $ab - bc = -ac$
    $b(a - c) = -ac$
        $b = -\dfrac{ac}{a - c}$
        $= \dfrac{ac}{c - a}$

   PHOTOCOPYING OF THIS PAGE IS RESTRICTED UNDER LAW.   ISBN: 9780170477680

**8**
$$\frac{4ab}{a} = 3ab - \frac{2ab}{b}$$
$$4b = 3ab - 2a$$
$$4b - 3ab = -2a$$
$$b(4 - 3a) = -2a$$
$$b = -\frac{2a}{4 - 3a}$$

**9** $a = \dfrac{bc}{d - c}$     **10** $b = \dfrac{c + 7a}{3a + 1}$

## Geometry and space (pp. 100–129)

### Right-angled triangles (pp. 102–118)

Theorem of Pythagoras (pp. 102–103)

**1** 8.6 cm     **2** 12.5 m
**3** 21.5 mm     **4** 1.75 km
**5** 34.1 cm     **6** 2.98 km
**7** $a = \sqrt{1.92 - 0.95^2} = 1.65$ m
**8** $b = 2\sqrt{38^2 - 35^2} = 29.6$ cm
**9** $AD = 6.35 - 3.75 = 2.6$ m
$BD = \sqrt{3.53^2 - 2.6^2} = 2.388$ m
$BC = \sqrt{2.388^2 + 3.75^2} = 4.446$ m
**10** $CD = \sqrt{0.47 - 0.38^2} = 0.277$ km
$BD = 0.277 + 0.21 = 0.487$
$AB = \sqrt{0.487^2 + 0.38^2} = 0.618$ km
**11** Hypotenuse $= \sqrt{x^2 + (x + 4)^2}$
$= \sqrt{x^2 + x^2 + 8x + 16}$
$= \sqrt{2x^2 + 8x + 16}$
**12** Hypotenuse $= \sqrt{(x - 3)^2 + (x + 2)^2}$
$= \sqrt{x^2 - 6x + 9 + x^2 + 6x + 9}$
$= \sqrt{2x^2 + 18}$
**13** Hypotenuse $= \sqrt{(4x + 1)^2 + (x + 2)^2}$
$= \sqrt{16x^2 + 8x + 1 - x^2 - 4x - 4}$
$= \sqrt{15x^2 + 4x - 3}$
**14** $s^2 = r^2 + (3r)^2$
$s^2 = 10r^2$
$s = \sqrt{10r^2}$ or $\sqrt{10}\,r$
**15** $x^2 + (2x - 2)^2 = (2x + 2)^2$
$x^2 + 4x^2 - 8x + 4 = 4x^2 + 8x + 4$
$x^2 + 16x = 0$
$x(x + 16) = 0$
The value of $x$ cannot be 0, so $x = 16$.
Sides are 16, 30 and 34.
**16** $x^2 + (x + 7)^2 = (x + 8)^2$
$x^2 + x^2 + 14x + 49 = x^2 + 16x + 64$
$x^2 - 2x - 15 = 0$
$(x - 5)(x + 3) = 0$
The value of $x$ cannot be −3, so $x = 5$.
Sides are 5, 12 and 13.
**17** $(3x)^2 + (3x + 4)^2 = (6x - 4)^2$
$9x^2 + 9x^2 + 24x + 16 = 36x^2 - 48x + 16$
$18x^2 - 72x = 0$
$18x(x - 4) = 0$
The value of $x$ cannot be 0, so $x = 4$.
Sides are 12, 16 and 20.

**18** $z^2 + (3z + 3)^2 = (2z + 11)^2$
$z^2 + 9z^2 + 18z + 9 = 4z^2 + 44z + 121$
$6z^2 - 26z - 112 = 0$
$2(3z^2 - 13z - 56) = 0$
$(3z + 8)(z - 7) = 0$
The value of $x$ cannot be $-\dfrac{8}{3}$, so $x = 7$.
Sides are 7, 24 and 25.
**19** DC is common to both triangles.
$\therefore \quad 103^2 - (120 - y)^2 = 70^2 - y^2$
$103^2 - 120^2 + 240y - y^2 = 70^2 - y^2$
$240y = 120^2 - 103^2 + 70$
$y = \dfrac{120^2 - 103^2 + 70}{240}$
$= 36.2$ cm
**20** $EG = \sqrt{(4x)^2 + (3x)^2}$
$= \sqrt{25x^2}$
$= 5x$
$EC = \sqrt{(5x)^2 + (12x)^2}$
$= \sqrt{169x^2}$
$= 13x$
$x = \dfrac{78}{13} = 6$
Dimensions of the cuboid are 24 cm by 18 cm by 72 cm.

Trigonometry (pp. 108–115)

**1** 2.95 m     **2** 33.4 cm
**3** 0.600 km     **4** 386 mm
**5** 941.5 cm     **6** 5.82 m
**7** 1.20 km     **8** 639.1 mm
**9** $x = 2(280 \cos 45°) = 396.0$ mm
**10** $z = 23 \sin 60° = 0.199$ m
**11** $x = 2(9.6 \cos 63°) = 8.72$ cm
**12** $y = \dfrac{0.046}{\cos 14°} = 0.0474$ km
**13** 50.4°     **14** 40.5°
**15** 38.5°     **16** 41.8°
**17** 61.2°     **18** 25.7°
**19** 30.5°     **20** 286.7°
**21** $AD = 2(33 \sin 32°) = 35.0$ cm
$x = \cos^{-1}\left(\dfrac{20}{32}\right) = 51°$
**22** $\angle ACB = \cos^{-1}\left(\dfrac{32}{36}\right) = 27°$
$\angle DCE = 45°$
$x = 20 \sin 45° = 14.1$ cm
**23** $\angle BAD = \sin^{-1}\left(\dfrac{2.9}{4.8}\right) = 37.2°$
$AD = 4.8 \cos 37.2° = 3.82$ cm
$\angle CAD = \tan^{-1}\left(\dfrac{1.7}{3.82}\right) = 24.0°$
$x = 37.2° - 24.0° = 13°$
**24** $BD = \dfrac{48}{\sin 56°} = 85.8$ cm
$y = \sin^{-1}\left(\dfrac{77}{85.8}\right) = 64°$

**25** AC = 36 sin 60° = 29.5 cm

$$CD = \frac{29.5}{\tan 35°} = 42.1 \text{ cm}$$

**26** x = 10 cos 37° = 8

y = 10 sin 37° = 6

Coordinates: (8, 0) (6, 0)

**Putting it together with Pythagoras (pp. 116–118)**

**1** 27.2 cm

**2** 0.566 m

**3** 36.9°

**4** 12.0 cm

**5** DE = 25 cos 59° = 12.88 cm

BD = 12.88 + 9 = 21.88 cm

$$x = \tan^{-1} \frac{21.88}{32} = 34.4°$$

**6** $\angle x = \sin^{-1} \frac{12}{21} = 34.8°$

AC = $\sqrt{21^2 - 12^2}$ = 17.23 cm, so DC = 8.617 cm

$$\angle y = \tan^{-1} \frac{8.617}{12} = 35.7°$$

**7** $AD = \frac{20}{\tan 35} = 28.56$ cm

CD = 28.56 tan 23° = 12.12 cm

BC = 20 – 12.12 = 7.9 cm (1 dp)

**8** MO = 16 sin 50° = 12.26 m

NM = 16 cos 50° = 10.28 m

MP = 10.28 tan 38° = 8.04 m

x = 12.26 – 8.04 = 4.22 m

**9**

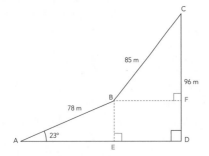

AE = 78 cos 23° = 71.80 m

BE = 78 sin 23° = 30.48 m

CF = 65.52 m

BF = $\sqrt{85^2 - 65.52^2}$ = 54.15 m

AD = 71.80 + 54.15 = 125.9 m

**10** $AD = \frac{x + 35}{\tan 66°}$ cm

$AD = \frac{x}{\tan 55°}$ cm

$\therefore \frac{x + 35}{\tan 66°} = \frac{x}{\tan 55°}$

x = 61.12 cm

**Challenge 4 (pp. 119–120)**

**1 a** tan⁻¹(smallest angle) = $\frac{1}{2}$

Smallest angle = 27°

**b** Ratio = $\sqrt{2^2 + 1^2}$:1 = 2.236:1

**2 a** sin (smallest angle) = $\frac{2}{7}$

Smallest angle = 17°

**b** Shortest side = 2 x 3 = 6 cm

Remaining side = $\sqrt{21^2 - 6^2}$ = 20.12 cm

**3** AB = $\sqrt{2x^2}$ = $\sqrt{2}$ x

AC = $\sqrt{\sqrt{(2x^2)^2 + x^2}}$ = $\sqrt{\sqrt{(2x^2 + x^2)}}$ = $\sqrt{3x^2}$ = $\sqrt{3}$ x

y = $\sqrt{\sqrt{(3x^2)^2 + x^2}}$ = $\sqrt{\sqrt{(3x^2 + x^2)}}$ = $\sqrt{4x^2}$ = 2x

**4 a** sin C = $\frac{h}{a}$ so h = a sin C

**b** Area = $\frac{1}{2}$ x b x a sin C

$= \frac{1}{2}$ a b sin C

**c** Traditional formula needs the lengths of the base and height.

New formula needs the lengths of two sides and the angle between them.

**5 a** By Pythagoras, BD = $\sqrt{2^2 - 1^2}$ = $\sqrt{3}$

**b**

| | Value | | Value |
|---|---|---|---|
| sin 60° | $\frac{\sqrt{3}}{2}$ | sin 30° | $\frac{1}{2}$ |
| cos 60° | $\frac{1}{2}$ | cos 30° | $\frac{\sqrt{3}}{2}$ |
| tan 60° | $\sqrt{3}$ | tan 30° | $\frac{1}{\sqrt{3}}$ |

**Pythagoras and trigonometry in 3D (pp. 121–124)**

**1 a** HC = $\sqrt{13^2 + 6^2}$ = 14.31 cm

**b** HB = $\sqrt{14.31^2 + 4^2}$ = 14.86 cm

**c** $\angle BHC = \tan^{-1}(\frac{4}{14.32})$ = 15.6°

**d** $\angle HFG = (\frac{4}{13})$ = 17.1°

**e** HA = $\sqrt{4^2 + 6^2}$ = 7.21 cm

$\angle BHA = \tan^{-1}(\frac{13}{7.21})$ = 61°

**f** BD = $\sqrt{4^2 + 13^2}$ = 13.60 cm

$\angle HBD = \tan^{-1}(\frac{6}{13.60})$ = 23.8°

**2 a** GC = HD = 75 tan 24° = 33.4 cm

**b** HC = $\sqrt{75^2 + 33.39^2}$ = 82.10 cm

HB = $\sqrt{82.10^2 + 56^2}$ = 99.38 cm

$\angle BHC = \cos^{-1}(\frac{82.10}{99.38})$ = 34.3°

**c** $\angle HFE = \tan^{-1}(\frac{56}{75})$ = 36.7°

**3 a** AC = $\sqrt{11^2 - 12^2}$ = 16.3 cm

**b** Angle = $\tan^{-1}(\frac{11}{12})$ = 42.5°

**4 a** VY = $\frac{\sqrt{3.2^2 + 3.2^2}}{2}$ = 2.262 m

Height = $\sqrt{5^2 - 2.262^2}$ = 4.5 m

**b** $\angle SUY = \cos^{-1}(\frac{2.262}{5})$ = 63.1°

PHOTOCOPYING OF THIS PAGE IS RESTRICTED UNDER LAW.  ISBN: 9780170477680

**c**  Angle $= \tan^{-1}\left(\frac{4.46}{1.6}\right) = 70.3°$

**5 a**  $DC = \sqrt{29^2 + 46^2} = 54.4$ cm

**b**  $GD = \sqrt{29^2 - \left(\frac{31}{2}\right)^2} = 24.51$ cm

$GA = \sqrt{46^2 + 24.51^2} = 52.1$ cm

**c**  $\angle DFE = \cos^{-1}\left(\frac{15.5}{29}\right) = 57.7°$

## Similar triangles (pp. 125–129)

Justification of similar triangles using angles (p. 126)

**1**  $\angle C = 98°$ ($\angle$s in $\Delta = 180°$)
∴ There are not two equal angles in each triangle.
∴ $\Delta ABC$ is not similar to $\Delta PRQ$.

**2**  $\angle A = 71°$ ($\angle$s in $\Delta = 180°$)
∴ Two angles in each triangle are equal.
∴ $\Delta ABC$ is similar to $\Delta FED$.

**3**  $\angle B = 80°$ ($\angle$s in $\Delta = 180°$)
∴ Two angles in each triangle are equal.
∴ $\Delta ABC$ is similar to $\Delta DFE$.

Justification of similar triangles using sides (pp. 127–129)

**1**  Equal sides: $\frac{PQ}{AC} = \frac{54}{72} = 0.75$

Short sides: $\frac{QR}{AB} = \frac{45}{60} = 0.75$

∴ $\Delta ABC$ and $\Delta QRP$ are similar because all their sides are proportional.
Scale factor = 0.75

**2**  Longest sides: $\frac{BC}{PQ} = \frac{84}{100} = 0.84$

Shortest sides: $\frac{AB}{RQ} = \frac{40}{48} = 0.8\dot{3}$

∴ $\Delta ABC$ and $\Delta RQP$ are not similar because not all their sides are proportional.

**3**  Longest sides: $\frac{PR}{ST} = \frac{60}{45} = 1.\dot{3}$

Middle sides: $\frac{PQ}{UT} = \frac{52}{39} = 1.\dot{3}$

Shortest sides: $\frac{QR}{SU} = \frac{32}{24} = 1.\dot{3}$

∴ $\Delta STU$ and $\Delta QRP$ are similar because all their sides are proportional.
Scale factor = $1.\dot{3}$

**4**  $\frac{AB}{ED} = \frac{AC}{x}$

$\frac{7.2}{6} = \frac{6.6}{x}$

$x = 5.5$ cm

**5**  $\frac{AE}{DC} = \frac{72}{36} = 2$

$x = 41 \times 2 = 82$ mm

$y = \frac{62}{2} = 31$ mm

**6**  $\frac{AB}{DF} = \frac{35}{42} = 0.8\dot{3}$

$x = 36 \times 0.8\dot{3} = 30$ mm

$y = \frac{22}{0.8\dot{3}} = 26.4$ mm

**7**  $\frac{AC}{BC} = \frac{10.35}{13.8} = 0.75$

$x = \frac{3.6}{0.75} = 4.8$ cm

$y = 12.2 - 12.2 \times 0.75 = 3.05$ cm

**8**  $\frac{BC}{GE} = \frac{68}{102} = 0.\dot{6}$

$x = 60 \times 0.\dot{6} = 20$ cm

$y = \frac{45}{0.\dot{6}} = 60$ cm

$z = \frac{60}{0.\dot{6}} = 90$ cm

**9**  $\frac{AB}{GE} = \frac{11}{4.4} = 2.5$

$AD = 2.88 \times 2.5 = 7.2$ cm

$FC = \frac{5.8}{2.5} = 2.32$ cm

**10**  $\Delta$s CDE and CBA are similar because AB and DE are similar.

$\frac{ED}{AB} = \frac{CE}{CA}$

$\frac{2.4}{3.6} = \frac{5.2}{CA}$

$CA = 7.8$ cm
$AE = 2.6$ cm

**11**  $\frac{AF}{AC} = \frac{FE}{AF}$

$\frac{3}{5} = \frac{FE}{3}$

$FE = \frac{9}{5} = 1.8$

∴ Area $= 3 \times 1.8$
$= 5.4$ m²

## Measurement (pp. 130–146)

### Three-dimensional shapes: surface area and volume (pp. 130–144)

Surface area (pp. 131–133)

**1**  SA $= (1.6 \times 0.5) + (1.6 \times 1.2) + (1.6 \times 1.3) + 2(0.5 \times 1.2) \times 0.5$
$= 5.4$ m²

**2**  SA $= 2\pi rh + 2\pi r^2$
$= 2\pi \times 47 \times 102 + 2\pi \times 47^2$
$= 44\,001$ cm² (5 sf) or $44\,000$ cm² (3 sf)

**3**  SA = area base + 4(area side)
$= 1.9^2 + 4\left(\frac{1}{2} \times 1.9 \times 2.4\right)$
$= 12.73$ m² (4 sf)

**4**  SA $= 2\left(\frac{95 + 192}{2} \times 94\right) + (134 \times 192) + 2(134 \times 106)$
$+ (134 \times 95)$
$= 93\,844$ mm² (0 dp)

**5**  SA $= 4\pi(0.35)^2$
$= 1.539$ m³ (4 sf)

**6**  Slant height ($l$) $= \sqrt{48^2 + 13^2}$
$= 49.729$ cm
SA $= \pi \times 13 \times 49.729 + \pi \times 13^2$
$= 2562$ cm² (0 dp)

**7** $SA = \frac{1}{2}(4\pi r^2) + \pi r^2$

$= 3 \times \pi \times 1.6^2$

$= 24.13 \text{ m}^2 \text{ (4 sf)}$

**8** Vertical height $= \sqrt{6^2 - 3^2}$

$= \sqrt{27}$

$SA = 4(\frac{1}{2} \times 6 \times \sqrt{27})$

$= 62.35 \text{ cm}^2 \text{ (4 sf)}$

**9** $SA = 2\pi rh + \pi r^2 + \frac{1}{2}(4\pi r^2)$

$= 2 \times \pi \times 9 \times 15 + 3 \times \pi \times 9^2$

$= 1612 \text{ cm}^2 \text{ (4 sf)}$

Volume (pp. 134–136)

**1** $V = \pi \times (0.65)^2 \times 2.9$

$= 3.849 \text{ m}^3 \text{ (4 sf)}$

**2** $V = \frac{1}{2} \times 16 \times 14 \times 20$

$= 2240 \text{ m}^3 \text{ (0 dp)}$

**3** $V = \frac{10 + 25}{2} \times 18 \times 90$

$= 28\,350 \text{ cm}^3 \text{ (0 dp)}$

**4** $V = (36 \times 56 - 28 \times 18) \times 70$

$= 105\,840 \text{ cm}^3 \text{ (0 dp)}$

**5** $V = \frac{1}{3}(2.3^2 \times 2.5)$

$= 4.408 \text{ m}^3 \text{ (4 sf)}$

**6** $V = \frac{4}{3} \times \pi \times 9.9^3$

$= 4064 \text{ mm}^3 \text{ (4 sf)}$

**7** $V = \frac{1}{3} \times \pi \times 35^2 \times 52$

$= 66\,706 \text{ cm}^3 \text{ (0 dp)}$

**8** $V = \frac{1}{3}\left(\frac{1}{2}(1.9 \times 2) \times 2.3\right)$

$= 1.457 \text{ m}^3 \text{ (4 sf)}$

**9** $V = \frac{4}{3} \times \pi \times 56^3$

$= 735\,619 \text{ cm}^3 \text{ (0 dp) or } 735\,600 \text{ (4 sf)}$

**10** $V = \frac{1}{3} \times \pi \times 5.1^2 \times 9.8$

$= 266.9 \text{ cm}^3 \text{ (4 sf)}$

**11** $V = \frac{1}{2}\left(\frac{4}{3} \times \pi \times 1.62^3\right)$

$= 8.904 \text{ m}^3 \text{ (4 sf)}$

**12** $V = \frac{3}{4}\left(\frac{4}{3} \times \pi \times 14.05^3\right)$

$= 8713 \text{ m}^3 \text{ (0 dp)}$

Areas and volumes with ratios (pp. 137–138)

**1** $1 \times 3^3 = 27 \text{ L}$

**2** $480 \times \left(\frac{1}{2}\right)^3 = 60 \text{ cm}^3$

**3** $10 \times (2.5)^2 = 62.5 \text{ cm}^2$

**4** $240 \times \left(\frac{3}{4}\right)^3 = 101.25 \text{ m}^3$

**5** $195 \times \left(\frac{2}{5}\right)^2 = 31.2 \text{ m}^3$

**6** $5yz \times 3^2 = 45xy$

**7** $1 \times \left(\frac{3}{2}\right)^2 = 3.375 \text{ L}$

**8** $1 : \sqrt[3]{\frac{192}{3}} = 1:4$

**9** $1 : \sqrt{\frac{67}{1675}} = 1:\frac{1}{5} = 5:1$

**10** $1 : \sqrt[3]{\frac{137.2}{50}} = 1:1.4 = 5:7$

Compound shapes: surface area and volume (pp. 139–141)

**1** SA of cone $= \pi rl$

$= \pi \times 0.8 \times \sqrt{0.8^2 + 2.1^2}$

$= 5.648 \text{ m}^2$

SA of cylinder $= \pi r^2 + 2\pi rh$

$= \pi \times 0.8^2 + 2\pi \times 0.8 \times 2.3$

$= 13.572 \text{ m}^2$

Total SA $= 19.22 \text{ m}^2 \text{ (2 dp)}$

Volume of cone $= \frac{1}{3}\pi r^2 \times h$

$= \frac{1}{3} \times \pi \times 0.8^2 \times 2.1$

$= 1.407 \text{ m}^3$

Volume of cylinder $= \pi r^2 \times h$

$= \pi \times 0.8^2 \times 2.3$

$= 4.624 \text{ m}^3$

Total volume $= 6.03 \text{ m}^3$

**2** SA hemisphere $= \frac{1}{2}(4\pi r^2)$

$= 2 \times \pi \times 0.28^2$

$= 0.4926 \text{ m}^2$

SA of cuboid $= 2(0.71 \times 0.90) + 2(0.71 \times 0.84 + 2(0.84 \times 090) - (\pi \times 0.28^2)$

$= 3.7365 \text{ m}^2$

Total SA $= 4.229 \text{ m}^2 \text{ (4 sf)}$

Volume of hemisphere $= \frac{1}{2}\left(\frac{4}{3}\pi r^2\right)$

$= \frac{4}{6} \times \pi \times 0.28^3$

$= 0.0460 \text{ m}^2$

Volume of cuboid $= 0.9 \times 0.71 \times 0.84$

$= 0.5368 \text{ m}^2$

Total volume $= 0.583 \text{ m}^2 \text{ (3 dp)}$

**3** $SA = 2(56 \times 24 + 37 \times 24 + 56 \times 37) - 2\left(\frac{1}{2} \times 19 \times 16.5\right) + 3(19 \times 24)$

$= 9662.5 \text{ cm}^2$

Volume $= (56 \times 37 \times 24) - \frac{1}{2} \times 19 \times 16.5 \times 24$

$= 45\,966 \text{ cm}^2$

**4** $SA = \pi \times 7 \times 24 + \pi \times 5 \times 24 + 2(\pi \times 3.5^2 - \pi \times 2.5^2)$

$= 168\pi + 120\pi + 12\pi$

$= 300\pi \text{ or } 942.5 \text{ cm}^2 \text{ (4 sf)}$

Volume $= \pi \times 3.5^2 \times 24 - \pi \times 2.5^2 \times 24$

$= 294\pi - 150\pi$

$= 144\pi \text{ or } 452.4 \text{ cm}^3 \text{ (4 sf)}$

PHOTOCOPYING OF THIS PAGE IS RESTRICTED UNDER LAW.    ISBN: 9780170477680

**5**  SA = 2(5.5 x 1.5 + 3 x 1.5 + 3 x 5.5) − ($\pi$ x 1$^2$) +

$\frac{1}{2}$($\pi$ x 2 x 5.5) − 2 x 5.5

= 58.5 − $\pi$ + 5.5$\pi$ − 11

= 61.64 m$^2$ (4 sf)

Volume = 3 x 5.5 x 1.5 − $\frac{1}{2}$ ($\pi$ x 1$^2$ x 5.5)

= 24.75 − 2.75$\pi$

= 16.11 m$^3$ (4 sf)

**6**  SA = $\pi$ x 53$^2$ + $\pi$ x 106 x 145 + $\pi$ x 53 x $\sqrt{53^2 + 140^2}$

= 2809$\pi$ + 15 370$\pi$ + 24 925.10

= 82 040 mm$^2$ (0 dp)

Volume = $\pi$ x 53$^2$ x 145 − $\frac{1}{3}$ x $\pi$ x 53$^2$ x 140

= 1 279 586 − 411 821

= 867 765mm$^3$ (0 dp)

**7**  Full height = 1.3 ÷ $\frac{1}{3}$

= 1.95 m

SA of whole cone = ($\pi r l$ + $\pi r^2$)

= $\pi$ x 1.02 x $\sqrt{1.02^2 + 1.95^2}$ + $\pi$ x 1.02$^2$

= 3.2851$\pi$

SA of small missing cone = $\pi r l$

= $\pi$ x 0.34 x $\sqrt{0.34^2 + 0.65^2}$

= 0.2494080191$\pi$

Bottom circle = $\pi r^2$

= $\pi$ x 0.34$^2$

= 0.1156$\pi$

Total surface area = 3.2851$\pi$ − 0.2494$\pi$

+ 0.1156$\pi$

= 3.1513$\pi$ or 9.900 m$^2$ (4 sf)

Volume = $\frac{1}{3}\pi r^2$ x 1.95 − $\frac{1}{3}\pi r^2$ x 0.65

= $\left(\frac{1}{3}$ x $\pi$ x 1.02$^2$ x 1.95$\right)$ − $\left(\frac{1}{3}$ x $\pi$ x 0.34$^2$ x 0.65$\right)$

= 2.0829 − 0.07867

= 2.004 m$^3$ (4 sf)

**Working backwards (pp. 142–143)**

**1**  V = 9.31 = $\pi$ x $r^2$ x 4.1

$r = \sqrt{\dfrac{9.31}{4.1\pi}}$

r = ±0.8502, but must be positive.

d = 1.700 cm (4 sf)

**2**  SA = 14 103 = 4 x $\pi$ x $r^2$

$r = \sqrt{\dfrac{14\ 103}{4\pi}}$

r = ±33.50

r must be positive so r = 33.50 cm (4 sf)

**3**  SA = 48 = 4$\left(\frac{1}{2}$ x b x 4$\right)$

b = 6 cm

**4**  SA = $\dfrac{4\pi r^2}{2}$ + $\pi r^2$

235.6 = 3$\pi r^2$

$r = \sqrt{\dfrac{235.6}{3\pi}}$

r = ±5.000 but cannot be negative.

r = 5.000 cm (4 sf)

**5**  V = ($\pi$ x(r + 2)$^2$ x 20) − ($\pi$ x $r^2$ x 20)

4800$\pi$ = 20$\pi$ (($r^2$ + 4r + 4) − $r^2$)

24 = 4r + 4

r = 5 cm

**6**  V = $\frac{1}{3}$ x $b^2$ x h + $b^2$ x h

$\frac{1}{3}$ x $b^2$ x 6 + $b^2$ x 8 = 490

10$b^2$ = 490

$b^2$ = $\dfrac{490}{10}$

b = $\sqrt{49}$

b = ±7

b must be positive so b = 7

**7**  V = $\frac{1}{3}$($b^2$ x 12) − $\frac{1}{3}$((3)$^2$ x 5)

241 = 4$b^2$ − 15

$b^2$ = 64

b = ± 8 but must be positive.

b = 8 m

**8**  SA = $\pi r^2$ + $\pi$ x r x 8 = 48$\pi$

48$\pi$ = $\pi r^2$ + 8$\pi r$

$r^2$ + 8r = 48

$r^2$ + 8r − 48 = 0

(r + 12)(r − 4) = 0

r = −12 or 4

The radius cannot be negative, so r = 4 cm

## Challenge 5 (pp. 144–146)

**1**  SA = 2($\frac{1}{2}$ x 12 x 8) + (15 x 12) + 2(15 x $\sqrt{6^2 + 8^2}$)

= 576 cm$^2$

**2**  SA = $\frac{1}{2}$(x + 3)(x + 4) x 4 + (x + 3)(x + 3)

= 2($x^2$ + 7x + 12) + $x^2$ + 6x + 9

= 3$x^2$ + 20x + 33 or (x + 3)(3x + 1)

**3**  SA = 2(y(y + 7) + y(y + 4) + y(y + 7) + y(y + 4))

= 2($y^2$ + 7y + $y^2$ + 4y + $y^2$ + 11y + 28)

= 6$y^2$ + 44y + 56

**4**  Slant height = $\sqrt{6^2 − 3^2}$

= $\sqrt{27}$

Vertical height = $\sqrt{27 − 9}$

= $\sqrt{18}$

Volume = $\frac{1}{3}$ x 6$^2$ x $\sqrt{18}$

= 50.91 m$^3$ (2 dp)

**5**
$$SA = \pi r^2 + \pi r l$$
$$28\pi = \pi r^2 + \pi \times r \times 3$$
$$28 = r^2 + 3r$$
$$r^2 + 3r - 28 = 0$$
$$(r + 7)(r - 4) = 0$$
$$r = -7 \text{ or } 4$$
$r$ cannot be negative so $r = 4$.

**6** SA of cube = SA of cube
$$6y^2 = 4\pi \times 9^2$$
$$6y^2 = 324\pi$$
$$y^2 = 54\pi$$
$$y = \sqrt{54\pi}$$
$$\therefore a = 54$$

**7** Height of hemisphere $= \dfrac{3}{5} \times 15$
$$= 9 \text{ cm}$$

Volume of shape $= \dfrac{1}{2}\left(\dfrac{1}{3} \times \pi r^3\right) + \dfrac{1}{3} \times r^2 \times h$

$$= \dfrac{2}{3} \times \pi \times 9^3 + \dfrac{1}{3} \times 9^2 \times 15$$

$$= 486\pi + 405\pi$$

$$= 891\pi \text{ or } 2799 \text{ cm}^3 \text{ (4 sf)}$$

**8** Volume of cylinder $= \pi r^2 \times h$
$$= \pi \times 2.5^2 \times 10$$
$$= 62.5\pi$$

Volume of spheres $= 2\left(\dfrac{4}{3} \times \pi r^3\right)$

$$= \dfrac{8}{3} \times \pi \times 2.5^3$$

$$= 41.\dot{6}\pi$$

Percentage $= \dfrac{41.\dot{6}}{62.5} \times 100 = 66.\dot{6}\%$

**9** $2 \times \dfrac{1}{3}\pi y^2 h = \dfrac{2}{3} \times \pi \times (4y)^3$ **(x 3)**
$$2\pi y^2 h = 2\pi \times 64y^3 \qquad (\div\, 2\pi)$$
$$y^2 h = 64y^3$$
$$h = \dfrac{64y^3}{y^2}$$
$$h = 64y$$

**10** Let $R$ be the radius of the cylinder and $r$ be the radius of the cone.

$$\pi R^2 h = 6 \times \dfrac{1}{3}\pi r^2 h \qquad (\div\, \pi \text{ and } h)$$
$$R^2 = 2r^2$$
$$R = \sqrt{2}r$$

∴ Radius of the cylinder is $\sqrt{2}$ times the radius of the cone.

## Proof (pp. 147–152)

### Geometric proof (pp. 147–149)

**1** $\angle q = \angle BCG$ (alt $\angle$s =, // lines)
$\angle t = \angle GCE$ (alt $\angle$s =, // lines)
$\angle r = \angle BCG + \angle GCE$
$\therefore \angle r = \angle q + \angle t$

**2** $\angle ADE = \angle ABC$ (corr $\angle$s =, // lines)
$\angle AED = \angle ACB$ (corr $\angle$s =, // lines)
$\angle A$ is common to both triangles
∴ Triangles ABC and ADE have equal angles.
∴ Triangles ABC and ADE are similar.

**3** $\angle ABC = 180° - \angle DBA - \angle CBE$ ($\angle$s on a line $\Rightarrow$ 180°)
But $\angle DBA = \angle BAC$ (alt $\angle$s =, // lines)
and $\angle CBE = \angle BCA$ (alt $\angle$s =, // lines)
$\therefore \angle ABC = 180° - \angle BAC - \angle BCA$
So $\angle ABC + \angle BAC + \angle BCA = 180°$

**4** $\angle BFE = \angle CBF$ (alt $\angle$s =, // lines)
∴ Two angles of the triangle are equal,
so BE = EF, and the triangle must be isosceles.

### Algebraic proof (pp. 150–152)

**1** Let the two odd numbers be $2x + 1$ and $2y + 1$.
Difference $= (2x + 1) - (2y + 1)$
$$= 2x + 1 - 2y - 1$$
$$= 2x - 2y$$
$$= 2(x - y)$$
$(x - y)$ must be an integer.
$2(x - y)$ must be even.
∴ The difference between two odd numbers must be even.

**2** Let the two consecutive integers be $n$ and $(n + 1)$.
Difference $= (n + 1)^2 - n^2$
$$= n^2 + 2n + 1 - n^2$$
$$= 2n + 1$$
$$= n + (n + 1)$$
$$= \text{the sum of the two consecutive integers.}$$

**3** $(3n + 1)^2 + (n - 1)^2 = 9n^2 + 6n + 1 + n^2 - 2n + 1$
$$= 10n^2 + 4n + 2$$
$$= 2(5n^2 + 2n + 1)$$
2(any integer) must be even.

**4** Let the three consecutive integers be $n$, $(n + 1)$ and $(n + 2)$.
Difference $= (n + 2)^2 - n^2$
$$= n^2 + 4n + 4 - n^2$$
$$= 4n + 4$$
$$= 4(n + 1)$$
$$= 4 \text{ times the middle number.}$$

**5** $(n + 3)(2n + 1) + (n - 2)(2n + 1) = 2n^2 + 7n + 3 +$
$$2n^2 - 3n - 2$$
$$= 4n^2 + 4n + 1$$
$$= 2(2n^2 + 2n) + 1$$
$2(2n^2 + 2n)$ must be even, so $2(2n^2 + 2n) + 1$ cannot be even.

**6** $(2n + 9)^2 - (2n + 5)^2 = 4n^2 + 36n + 81 - 4n^2 - 20n - 25$
$$= 16n + 56$$
$$= 4(4n + 14)$$
4(any integer) must be a multiple of four.

PHOTOCOPYING OF THIS PAGE IS RESTRICTED UNDER LAW. ISBN: 9780170477680

**7** $a + b = 9$

$\therefore b = 9 - a$

The two-digit number $= 10a + b$

$$= 10a + (9 - a)$$
$$= 10a + 9 - a$$
$$= 9a - 9$$
$$= 9(a - 1)$$

$9(a - 1)$ is always a multiple of 9.

**8** $(n + 1)^3 - (n + 1)^2 = (n + 1)(n^2 + 2n + 1) - (n^2 + 2n + 1)$

$$= (n + 1)(n^2 + 2n + 1) - n^2 - 2n - 1$$
$$= n^3 + 2n^2 + n + n^2 + 2n + 1 - n^2 - 2n - 1$$
$$= n^3 + 2n^2 + n$$
$$= n(n^2 + 2n + 1)$$

$n$(any integer) must be a multiple of $n$.

## Practice sets (pp. 153–164)
### Practice set one (pp. 153–154)

**1** $\dfrac{216a^6}{6a^7} = \dfrac{36}{a}$

**2** $-8x + 4x^2 - 4xy + 3x = -5x + 4x^2 - 4xy$

**3** $\dfrac{-2 \times 7 - 7 \times 4}{4^2} = -\dfrac{42}{16}$

$$= -\dfrac{21}{8} \text{ or } -2.625$$

**4** $A(x - 2y) = (4x - 3y)$

$Ax - 2Ay = 4x - 3y$

$Ax - 4x = 2Ay - 3y$

$x(A - 4) = 2Ay - 3y$

$$x = \dfrac{2Ay - 3y}{A - 4}$$

**5** $3x(5x + 7) - 4(2x - 1) = 15x^2 + 13x + 4$

**6** $x^2 + (2x + 4)2 = (3x - 4)^2$

$5x^2 + 16x + 16 = 9x^2 - 24x + 16$

$4x^2 - 40x = 0$

$4x(x - 10) = 0$

$x = 0 \text{ or } 10$

If $x = 0$ there is no triangle, so sides are 10, 24 and 26.

**7** $\quad 2^{x^2} > 2^x \times 16^3$

$\quad 2^{x^2} > 2^x \times (2^4)^3$

$\quad 2^{x^2} > 2^x \times 2^{12}$

$x^2 - x - 12 > 0$

$(x - 4)(x + 3) > 0$

$\quad x < -4 \text{ or } x > 3$

**8** $4(x^2 + 9x + 14) = 4(x + 7)(x + 2)$

**9** $y = a(x + 4)^2 - 2$

Graph passes through $(-2, 2)$.

$\therefore 2 = a(-2 + 4)^2 - 2$

$a = 1$

$y = (x + 4)^2 - 2$ or $x^2 + 8x + 14$

**10** $x + (2x + 1) + (3x - 1) + (6x + 2) + (2x - 1) + (4x + 1)$

$= 18x + 2$

**11** $V = \dfrac{4}{3} \times \pi \times r^3$

$288\pi = \dfrac{4}{3} \times \pi \times r^3$

$216 = r^3$

$r = \sqrt[3]{216}$

$r = 6$

**12** $\angle BCA$:

$\cos BCA = \dfrac{3.5}{43}$

$\angle BCA = 35.52° \text{ (2 dp)}$

$\angle DCE = 180 - 115 - 35.52$

$= 29.48° \text{ (2 dp)}$

Side CE:

$\sin 29.48° = \dfrac{2}{CE}$

$CE = 4.06 \text{ m (2 dp)}$

### Practice set two (pp. 155–156)

**1** $\dfrac{5x(x + 5)}{(x + 5)(x - 5)} = \dfrac{5x}{x - 5}$

**2** $5^2 - 4x = 6\cos 60°$

$25 - 4x = 3$

$-4x = -22$

$x = 5.5$

**3** $-5x^2 - 2xy + 3x - 4x + 24y - 28$

$= -5x^2 - 2xy - x + 24y - 28$

**4** $3(x + 6) = 54$

$x = 12$

**5** $12 \times 9 \times 7 : \dfrac{1}{2} \times 8 \times 5 \times 7$

$756 \text{ cm}^3 : 140 \text{ cm}^3$

$27 : 5$

**6** Circumference $= 2\pi r = 18\pi$

$r = 9 \text{ cm}$

Angle $= \tan^{-1}\left(\dfrac{10}{9}\right) = 48.0°$

**7** $\quad 2^{x^2 - 1} = 2^5 \div 2^{3x}$

$\quad 2^{x^2 - 1} = 2^{5 - 3x}$

$x^2 + 3x - 4 = 0$

$(x + 4)(x - 1) = 0$

$\quad x = -4 \text{ or } x = 1$

**8** $16x + 20y = 108$

$25x - 20y = 220$

$41x = 328$

$x = 8$

$y = -1$

$(8, -1)$

**9** $(3x - 2) + (4x + 3) + (2x + 1) + (5x + 2) + FG$

$= 15x + 10$

$14x + 4 + FG = 15x + 10$

$FG = x + 6$

**10** $y = -a(x - 4)^2 + 7$

Passes through $(0, -9)$

$-9 = -a(0 - 4)^2 + 7$

$-9 = -16a + 7$

$\therefore a = 1$

$y = -(x - 4)^2 + 7$ or $y = -x^2 + 8x - 9$

**11**  $5(x^2 - 9) = 5(x + 3)(x - 3)$

**12**  $\dfrac{6x - 2}{3} + \dfrac{3x - 1}{4} = \dfrac{4(6x - 2)}{12} + \dfrac{3(3x - 1)}{12}$

$= \dfrac{24x - 8 + 9x - 3}{12}$

$= \dfrac{33x - 11}{12}$

**13**  $7 \times 7 - 6.3 \times 6.3 \le \text{Area} \le 10 \times 10 - 5.1 \times 5.1$

$9.31 \text{ cm}^2 \le \text{Area} \le 73.99 \text{ cm}^2$

**Practice set three (pp. 157–158)**

**1**  $9(3x^2 - 10x + 63) = 9(x - 1)(3x - 7)$

**2**  $x > 10.5$

$x$ can be only integers $\Rightarrow x \ge 11$

**3**  $36 < x^2 < 144$

$6 < x < 12$ or $-6 < x < -12$

**4**  $6 + 2(3 \times (-3)^2) - (-4) + 7) = 6 + 2(38)$

$= 82$

**5**  $6^3 \div 0.6 = 9 \times 4 \times y$

$360 = 36y$

$y = 10 \text{ cm}$

**6**  Side BD:

$\tan 29° = \dfrac{BD}{30.9}$

$BD = 17.128$ (3 dp)

Area of $\triangle CBD = \dfrac{1}{2} \times 30.9 \times 17.128$

$= 264.63 \text{ cm}^2$ (2 dp)

Side AD:

$\tan 36° = \dfrac{17.13}{AD}$

$AD = 23.575$ (3 dp)

Area of $\triangle ADB = \dfrac{1}{2} \times 23.575 \times 17.128$

$= 201.90 \text{ cm}^2$ (2 dp)

**7**  $\dfrac{3ab(2ac - 3b)}{3ab(5c + a - 4bc^2)} = \dfrac{2ac - 3b}{5c + a - 4bc^2}$

**8**  $4x^2 + 4x - 3 \ge 4x^2 - 10x - 6$

$4x - 3 \ge -10x - 6$

$14x \ge -3$

$x \ge \dfrac{-3}{14}$

**9**  Turning point at (1, 3) $\Rightarrow y = -a(x - 1)2 + 3$

Passes through (4, –6) $\Rightarrow -6 = -a(4 - 1)2 + 3$

so $a = 1$ and $y = -(x - 1)^2 + 3$

**10**

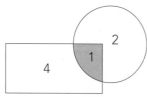

7 parts = 147 m$^2$

1 part = 21 m$^2$

Shaded area 21 m$^2$

Area rectangle = 5 parts

$= 105 \text{ m}^2$

**11**  $2y + 1 = 3x$ $\qquad$ (x 8)

$7y - 4 = 8x$ $\qquad$ (x –3)

$16y + 8 = 24x$

$-21y + 12 = -24x$

$-5y + 20 = 0$

$y = 4$

$8 + 1 = 3x$

$x = 3$

Sides are 24 and 9

**Practice set four (pp. 159–160)**

**1**  $2^8 = 256$, so $x = 8$.

**2**  $3(x + 3) - 2(2x - 1) = 6$

$3x + 9 - 4x + 2 = 6$

$-x + 11 = 6$

$-x = -5$

$x = 5$

**3**  $(x - 3)(x + 5) = 0$

$-5 < x < 3$

**4**  Area $= (3x + 2)(x - 4)$

$= 3x^2 - 10x - 8$

**5**  $\dfrac{32}{x} = \dfrac{24}{16}$

$x = \dfrac{32 \times 18}{24}$

$x = 24$

Perimeter of large = 112 cm

Perimeter of small = 84 cm

Ratio = 112:84 = 4:3

**6**  Slant height $= \sqrt{(9x)^2 + (12x)^2}$

$= 15x$

Curved SA $= \pi r l$

$\pi \times (9x) \times (15x) = 540 \pi$

$135x^2 \pi = 540 \pi$

$x^2 = 4$

$x = \pm 2$

$x$ cannot be negative so $x = 2$.

Volume $= \dfrac{1}{3}\pi r^2 h$

$= \dfrac{1}{3}\pi \times 18^2 \times 24$

$= 2592 \pi$

$a = 2592$

**7**  $x^2 - 6x + 5 = 0$

$(x - 1)(x - 5) = 0$

$x = 1$ or $5$

**8**  $x^2 - 3x - 10 > 0$

$(x - 5)(x + 2) > 0$

$x < -2$ or $x > 5$

PHOTOCOPYING OF THIS PAGE IS RESTRICTED UNDER LAW.   ISBN: 9780170477680

**9** Let $r$ = radius of sphere and $R$ = radius of cylinder.

$$\frac{4}{3} \times \pi r^3 = \pi R^2 \times h$$

$$\frac{4}{3} \times \pi r^3 = \pi(6x)^2 \times 5x$$

$$\frac{4}{3} \times \pi r^3 = 180\pi x^3$$

$$r^3 = 135x^3$$

$$r = \sqrt[3]{135}x$$

**10** $8x + 12y = 44$

$6x + 4y = 28$     (× –3)

$-18x - 12y = -84$

$-10x = -40$

$x = 4$ and $y = 1$

Rectangle A is 6 cm by 16 cm.

Rectangle B is 12 cm by 2 cm.

**11** $\frac{1}{2}(3x - 15)(2x - 4) = 210$

$(3x - 15)(x - 2) = 210$

$3x^2 - 21x - 180 = 0$

$x^2 - 7x - 60 = 0$

$(x - 12)(x + 5) = 0$

$x = 12$ or $-5$

$x$ cannot be negative so $x = 12$ cm

### Practice set five (pp. 161–162)

**1** $\dfrac{12x - 4}{24} + \dfrac{6x}{24} = \dfrac{18x - 4}{24}$

$$= \frac{9x - 2}{12}$$

**2** $\dfrac{(x - 4)(x + 3)}{(x + 3)(x - 9)} = \dfrac{x - 4}{x - 9}$

**3** $50 + 2x < 7x + 6$

$44 < 5x$

$x > 8.8$

Smallest number x can take is 11.

**4** Turning point: (3,–4)

This is a minimum.

**5** $2\pi r^2 + 2\pi rh = 130\pi$

$2\pi(r^2 + 8r) = 130\pi$

$r^2 + 8r - 65 = 0$

$(r + 13)(r - 5) = 0$

$r = -13$ or $5$

The radius cannot be negative, so $r = 5$ cm.

**6** Scale factor $= \dfrac{6}{4}$

$= 1.5$

$x = 6 \times 1.5$

$= 9\text{m}$

$y = 7.5 \div 1.5$

$= 5$ m

**7** $-3x + 7y = 1.125$    (× –8)

$-8x + 5y = 3$      (× 3)

$24x - 56y = -9$

$\underline{-24x + 15y = 9}$    (+)

$-41y = 0$

$(-0.375, 0)$

**8** $(x + 6)(x + 6) = 0$

$\therefore$ b = 12

**9** Horizontal asymptote is at $y = 3$, and the graph passes through (0, 4), so $y = 2^x + 3$.

**10** $ab^2 - 8c = 3d^2 + 12eb^2$

$ab^2 - 12eb^2 = 3d^2 + 8c$

$b^2(a - 12e) = 3d^2 + 8c$

$$b = \pm\sqrt{\frac{3d^2 + 8c}{a - 12e}}$$

**11** SA of sphere $= 4\pi r^2$

$= (4 \times \pi \times 6^2)$

$= 144\pi$

SA of cylinder $= 2\pi r^2 + 2\pi rh$

$= 8\pi + 4\pi h$

$8\pi + 4\pi h = \dfrac{144}{2}\pi$

$4\pi h = 64\pi$

$h = 16$ m

Volume sphere $= \dfrac{4}{3}\pi(6)^3 = 288\pi$

Volume cylinder $= \pi(2)^2 \times 16 = 64\pi$

Ratio $= 288\pi : 64\pi = 9:2$

**12** $\angle EAB = 180° - 76° - 59°$

$= 45°$

Height of $\triangle ABE$:

$\sin 45° = \dfrac{\text{height}}{4}$

Height $= 2.828$ m (3 dp)

Length of AE $= \cos 45° \times 4 + 2.828 \div \tan 59°$

$= 4.528$ m (3 dp)

Volume of prism $= \dfrac{1}{2} \times 4.528 \times 2.828 \times 3.35$

$= 21.45$ m$^3$ (2 dp)

### Practice set six (pp. 163–164)

**1**

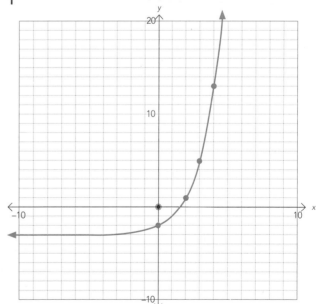

**2**    $m = \dfrac{-5 - 7}{12 - (-4)} = -\dfrac{3}{4}$

   $y - 7 = -\dfrac{3}{4}(x - (-4))$

   $y = -\dfrac{3}{4}x + 4$

**3**    $m_1 = -\dfrac{4}{5} \Rightarrow m_2 = \dfrac{5}{4}$

   $y - (-9) = \dfrac{5}{4}(x - 20)$

   $y = \dfrac{5}{4}x - 34$

**4**    $y = 3x^2 - 4x + 1$

**5**    $-22.5$

**6**         $(x + 3)^2 = 17^2 - 15^2$

   $x^2 + 6x - 55 = 0$

   $(x - 5)(x + 11) = 0$

   $x$ cannot be 11 because $x - 3$ cannot be negative

   Area = 120 cm²

**7**    $248 \times \left(\dfrac{7}{2}\right)^3 = 10\ 633$ cm³

**8**    $5^{2n + 1} = 5^{3n - 6}$

   $2n + 1 = 3n - 7$

   $n = 7$

**9**    $(-3, 9) \Rightarrow 9 = 9a - 3b - 6$

   $3a - b = 5$         ①

   $(2, 4) \Rightarrow 4 = 4a + 2b - 6$

   $2a + b = 5$         ②

   Add ① and ②:

   $5a = 10$

   $a = 2$

   Substitute for a in ②: $4 + b = 5$

   $b = 1$

   Equation: $2x^2 + x - 6$

**10**    $\dfrac{2}{3(15x^2 + x - 2)} \div \dfrac{1}{9x^2 - 1}$

   $= \dfrac{2}{3(3x - 1)(5x + 2)} \times \dfrac{(3x + 1)(3x - 1)}{x(x - 3)}$

   $= \dfrac{2(3x + 1)}{3(5x + 2)}$

   $= \dfrac{6x + 2}{15x + 6}$

   $a = 6$, $b = 2$, $c = 15$ and $d = 6$

PHOTOCOPYING OF THIS PAGE IS RESTRICTED UNDER LAW.    ISBN: 9780170477680